# High Performance with Java

Discover strategies and best practices to develop
high performance Java applications

**Dr. Edward Lavieri Jr.**

# High Performance with Java

**Group Product Manager**: Kunal Sawant
**Publishing Product Manager**: Teny Thomas
**Book Project Manager**: Manisha Singh
**Senior Editor**: Kinnari Chohan
**Technical Editor**: Vidhisha Patidar
**Copy Editor**: Safis Editing
**Proofreader**: Kinnari Chohan
**Indexer**: Subalakshmi Govindhan
**Production Designer**: Aparna Bhagat
**DevRel Marketing Coordinator**: Sonia Chauhan

First published: July 2024

Production reference: 1080724

Published by Packt Publishing Ltd.
Grosvenor House
11 St Paul's Square
Birmingham
B3 1RB, UK

ISBN 978-1-83546-973-6

www.packtpub.com

*To Brenda, my wonderful wife, whose unwavering support and patience during countless evenings and weekends made this book possible. And to KuBougie, our beloved Hemmingway cat, who brought joy and comfort to our home, helping to heal the loss of our dear dogs, Muzz and Java.*

*Thank you both!*

# Contributors

## About the author

**Dr. Edward Lavieri Jr.** is a seasoned software developer and senior member of the **Institute of Electrical and Electronic Engineers (IEEE)**. He has developed software for over 30 years and primarily focused on Java for over 20 years. His experience includes both military information systems and the software design and development industry. Dr. Lavieri currently serves as a university dean for a college of computer science and a school of information technology as well as the chief academic officer of a programming bootcamp organization, all part of a regionally accredited US university. He holds a doctorate degree in computer science and a master's degree in information systems, along with other degrees and certificates.

*I want to thank my wife, Brenda, for her unwavering support, and the entire Packt team, with special gratitude to Teny, Manisha, Kinnari, and Vidhisha.*

# About the reviewer

**Aristides Villarreal** is a Panamanian Java Champion and a member of the Apache Project Management Committee (PMC) for the NetBeans Project. He is also a Jakarta EE Ambassador and a specialist in NoSQL, Jakarta EE, and Eclipse MicroProfile.

**Pallavi Sharma** is a versatile professional with 18 years of experience, having worked as an individual contributor, project manager, Scrum Master, and coach. She is the Founder of 5 Elements Learning and the author of five books on Selenium. As a recognized committer to the Selenium Project, Pallavi actively participates in international testing conferences as a reviewer, judge, organizer, and speaker. She holds various certifications in her field. Outside of work, Pallavi enjoys writing, traveling, and nature watching. She is dedicated to giving back to society, guided by the principle of #bekind.

# Table of Contents

# 3

# Optimizing Loops                                          41

# 4

# Java Object Pooling                                       59

# 5

# Algorithm Efficiencies                                    69

# Part 2: Memory Optimization and I/O Operations

6

## Strategic Object Creation and Immutability        79

7

## String Objects                                     97

# 8

## Memory Leaks                                                                107

# Part 3: Concurrency and Networking

# 9

## Concurrency Strategies and Models                                          123

# 10

## Connection Pooling                                                          137

11

# Part 4: Frameworks, Libraries, and Profiling

12

13

# 14

# Profiling Tools                                        195

# Part 5: Advanced Topics

## 15

## 16

# Preface

Java remains one of the most popular and powerful programming languages in the world. Since its inception, it has empowered developers to create robust, high-performance applications across various domains. This book, *High Performance with Java*, delves into the intricate details of optimizing Java applications for peak performance. From understanding the fundamentals of the **Java virtual machine (JVM)** to leveraging advanced profiling tools, this book is a comprehensive guide for anyone looking to enhance their Java development skills and deliver high-performance solutions.

## Who this book is for

This book is intended for Java developers who have a foundational understanding of the language and are looking to deepen their knowledge of performance optimization. Whether you are a seasoned developer or a mid-level programmer, this book can provide valuable insights and practical techniques to improve the performance of your Java applications. A basic familiarity with Java development tools and the **Java Development Kit (JDK)** is assumed.

## What this book covers

*Chapter 1, Peeking Inside the Java Virtual Machine*, provides an in-depth look at the JVM, exploring how it works, including garbage collection and **just-in-time** (JIT) compiler optimizations to maximize application performance.

*Chapter 2, Data Structures*, allows you to learn about the impact of different data structures on performance and how to select and implement the most efficient ones for your applications.

*Chapter 3, Optimizing Loops*, covers techniques for optimizing loops, a fundamental construct in programming, to avoid bottlenecks and improve runtime performance.

*Chapter 4, Java Object Pooling*, dives into the concept of object pooling in Java and how to achieve high performance with them in your Java applications.

*Chapter 5, Algorithm Efficiencies*, helps you learn how to choose the right algorithm for any given requirement.

*Chapter 6, Strategic Object Creation and Immutability*, presents information and examples on how to make objects immutable and explains why you should consider it.

*Chapter 7, String Objects*, focuses on the efficient use of string objects in our Java applications.

*Chapter 8, Memory Leaks*, provides techniques, design patterns, coding examples, and best practices to avoid memory leaks.

*Chapter 9, Concurrency Strategies and Models*, provides foundational information on concurrency concepts and provides hands-on opportunities to implement concurrency in Java programs.

*Chapter 10, Connection Pooling*, covers the concept of connection pooling, providing fundamental principles, implementation approaches, and examples.

*Chapter 11, Hypertext Transfer Protocols*, covers the foundational protocol used for information exchange on the web.

*Chapter 12, Frameworks for Optimization*, provides a foundational understanding of key frameworks to optimize Java applications.

*Chapter 13, Performance-Focused Libraries*, takes an in-depth look at libraries such as JMH, Netty, and Jackson, and how they can be used to enhance the performance of your Java applications.

*Chapter 14, Profiling Tools*, explores various profiling tools, both bundled in the JDK and additional ones, to monitor and improve the performance of your Java applications.

*Chapter 15, Optimizing Your Databases with SQL Queries*, demonstrates how to optimize SQL queries and manage databases to ensure high performance.

*Chapter 16, Code Monitoring and Maintenance*, examines the importance of code monitoring and maintenance, including the use of **application performance management** (**APM**) tools, code reviews, and logging.

*Chapter 17, Unit and Performance Testing*, covers strategies for effective unit and performance testing to ensure your applications meet performance requirements.

*Chapter 18, Leveraging Artificial Intelligence (AI) for High-Performance Java Applications*, explores how AI can be integrated into Java applications for performance enhancements, including predictions, forecasting, and content pre-processing.

## To get the most out of this book

To fully benefit from the content in this book, readers should have a solid understanding of Java programming basics and be comfortable with development tools such as **integrated development environments** (**IDEs**) and version control systems. Familiarity with performance monitoring and profiling concepts will be advantageous. This book assumes that readers have a desire to delve deeper into the intricacies of Java performance tuning and are willing to experiment with different techniques to achieve the best results.

| Software/hardware covered in the book | OS requirements |
| --- | --- |
| Java 21 | Windows, macOS, or Linux |
| JDK 21.0.1 | |
| IDE | |

Libraries and frameworks are included throughout the book and their installation is optional.

**If you are using the digital version of this book, we advise you to type the code yourself or access the code from the book's GitHub repository (a link is available in the next section). Doing so will help you avoid any potential errors related to the copying and pasting of code.**

## Download the example code files

You can download the example code files for this book from GitHub at `https://github.com/PacktPublishing/High-Performance-with-Java`. If there's an update to the code, it will be updated in the GitHub repository.

We also have other code bundles from our rich catalog of books and videos available at `https://github.com/PacktPublishing/`. Check them out!

## Conventions used

There are a number of text conventions used throughout this book.

`Code in text`: Indicates code words in text, database table names, folder names, filenames, file extensions, pathnames, dummy URLs, user input, and Twitter handles. Here is an example: "As you can see in the following, we can use the `javap` command without any parameters."

A block of code is set as follows:

```
public class CH1EX1 {
  public CH1EX1 {
  public static void main(java.lang.String[]);
  }
}
```

When we wish to draw your attention to a particular part of a code block, the relevant lines or items are set in bold:

```
public class Example2 {
  public static void main(String[] args) {
    List<String> petNames = new LinkedList<>();
```

Any command-line input or output is written as follows:

```
$ cd src
$ ls
CH1EX1.java
$
```

> **Tips or important notes**
> Appear like this.

# Get in touch

Feedback from our readers is always welcome.

**General feedback**: If you have questions about any aspect of this book, email us at customercare@ packtpub.com and mention the book title in the subject of your message.

**Errata**: Although we have taken every care to ensure the accuracy of our content, mistakes do happen. If you have found a mistake in this book, we would be grateful if you would report this to us. Please visit www.packtpub.com/support/errata and fill in the form.

**Piracy**: If you come across any illegal copies of our works in any form on the internet, we would be grateful if you would provide us with the location address or website name. Please contact us at copyright@packt.com with a link to the material.

**If you are interested in becoming an author**: If there is a topic that you have expertise in and you are interested in either writing or contributing to a book, please visit authors.packtpub.com.

## Share Your Thoughts

Once you've read *High Performance with Java*, we'd love to hear your thoughts! Scan the QR code below to go straight to the Amazon review page for this book and share your feedback.

https://packt.link/r/1835469736

Your review is important to us and the tech community and will help us make sure we're delivering excellent quality content.

# Download a free PDF copy of this book

Thanks for purchasing this book!

Do you like to read on the go but are unable to carry your print books everywhere?

Is your eBook purchase not compatible with the device of your choice?

Don't worry, now with every Packt book you get a DRM-free PDF version of that book at no cost.

Read anywhere, any place, on any device. Search, copy, and paste code from your favorite technical books directly into your application.

The perks don't stop there, you can get exclusive access to discounts, newsletters, and great free content in your inbox daily

Follow these simple steps to get the benefits:

1.  Scan the QR code or visit the link below

https://packt.link/free-ebook/978-1-83546-973-6

2.  Submit your proof of purchase

3.  That's it! We'll send your free PDF and other benefits to your email directly

# Part 1:
# Code Optimization

This part delves into the core techniques and strategies to optimize Java code. It starts with an in-depth exploration of the **Java Virtual Machine (JVM)** and its impact on performance. You will learn about the efficient use of data structures, techniques to optimize loops, the benefits of Java object pooling, and strategies to improve algorithm efficiencies. By mastering these fundamental concepts, you will significantly enhance the performance of your Java applications.

This part has the following chapters:

- *Chapter 1, Peeking Inside the Java Virtual Machine (JVM)*
- *Chapter 2, Data Structures*
- *Chapter 3, Optimizing Loops*
- *Chapter 4, Java Object Pooling*
- *Chapter 5, Algorithm Efficiencies*

# 1

# Peeking Inside the Java Virtual Machine

I would like to introduce you to a remarkable piece of technology that helped revolutionize the software industry. Meet the **Java Virtual Machine** (**JVM**). Okay, so you are likely already familiar with the JVM, and it is important to understand and appreciate the tremendous value it has as the middleware between compiled Java bytecode and a virtually limitless number of hardware platforms. The ingenious design of Java and the JVM is a testament to its wild popularity and value to users and developers.

Since the first release of Java in the 1990s, the JVM has been the true success factor for the Java programming language. With Java and the JVM, the concept of "write once, run anywhere" was born. Java developers can write their programs once and allow the JVM to ensure the code runs on devices the JVM is installed on. The JVM also makes Java a platform-independent language. The primary objective of this chapter is to provide greater insights into the unsung hero of Java, the JVM.

In this chapter, we will take an extensive look at the JVM so that we can learn to get the most out of it in an effort to increase the performance of our Java applications.

In this chapter, we will cover the following topics:

- How the JVM works
- Garbage collection
- **Just-in-Time** (**JIT**) compiler optimizations

By the end of this chapter, you should understand how to get the most out of the JVM to improve the performance of your Java applications.

## Technical requirements

To follow the instructions in this chapter, you will need the following:

- A computer with either Windows, macOS, or Linux installed
- The current version of the Java SDK
- Preferably, a code editor or **integrated development environment** (IDE) (such as Visual Studio Code, NetBeans, Eclipse, or IntelliJ IDEA)

The finished code for this chapter can be found here: `https://github.com/PacktPublishing/High-Performance-with-Java/tree/main/Chapter01`.

> **Important note**
> This book is based on Java 21 and the **Java Development Kit** (JDK) 21.0.1. It also uses the IntelliJ IDEA Community Edition IDE, running on macOS.

This chapter contains code examples that you can use to follow along with and experiment with. So, you will want to ensure your system is properly prepared.

Start by downloading and installing your IDE of choice. Here are a couple of options:

- Visual Studio Code (`https://code.visualstudio.com/download`)
- NetBeans (`https://netbeans.apache.org/download/`)
- Eclipse (`https://www.eclipse.org/downloads/`)
- IntelliJ IDEA Community Edition (`https://www.jetbrains.com/idea/download/`)

After your IDE is set up, you need to ensure it is configured for Java development. Most modern IDEs are capable of downloading and installing the Java SDK for you. If this is not the case for you, the Java SDK can be obtained here: `https://www.oracle.com/java/technologies/downloads/`.

Ensuring that your IDE is Java-ready is important. Here are some IDE-specific links in case you need some help:

- Java in Visual Studio Code (`https://code.visualstudio.com/docs/languages/java`)
- Java quick start tutorial for NetBeans (`https://netbeans.apache.org/tutorial/main/kb/docs/java/quickstart/`)

- Preparing Eclipse (`https://help.eclipse.org/latest/index.jsp?topic=%2Forg.eclipse.jdt.doc.user%2FgettingStarted%2Fqs-3.htm`)

- Creating your first Java application with IntelliJ IDEA Community Edition (`https://www.jetbrains.com/help/idea/creating-and-running-your-first-java-application.html`)

Once you have your IDE and the Java SDK installed and configured on your computer, you are ready to move to the next section.

## How the JVM works

At the core, the JVM sits between your Java bytecode and your computer. As illustrated next, we develop our Java source code in an IDE, and our work is saved as `.java` files. We use the Java compiler to convert our Java source code to bytecode; the resulting files are `.class` files. We then use the JVM to run our bytecode on our computer:

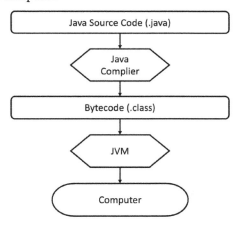

Figure 1.1 – Java application workflow

It is important to realize that Java is different from typical compile-and-execute languages where source code is fed into a compiler, which then produces a `.exe` file (for Windows), a `.app` file (for (macOS), or ELF files (for Linux). This impressive process is much more complex than it seems.

Let's take a closer look at what is going on with regard to the JVM, using a basic example. In the following code, we implement a simple loop that prints to the console. It is presented so that we can see how the JVM handles Java code:

```
// Chapter1
// Example 1
public class CH1EX1 {
```

```java
    public static void main(String[] args) {

        System.out.println("Basic Java loop.");

        // Basic loop example
        for (int i = 1; i <= 5; i++) {
            System.out.println("1 x " + i + " = " + i);
        }
    }
}
```

We can use the Java compiler to compile our code into a .class file, which will be formatted as bytecode. Let's open a terminal window, navigate to our project's src folder as shown next, and use the ls command to reveal our .java file:

```
$ cd src
$ ls
CH1EX1.java
$
```

Now that we are in the proper folder, we can use javac to convert our source code to bytecode. As you can see next, this created a .class file:

```
$ javac CH1EX1.java
$ ls
CH1EX1.class    CH1EX1.java
$
```

Next, we can use the **Java Printer** tool (javap) to print a decompiled version of our bytecode. As you can see next, we can use the javap command without any parameters:

```
javap CH1EX1.class
Compiled from "CH1EX1.java"
public class CH1EX1 {
  public CH1EX1();
  public static void main(java.lang.String[]);
}
```

As you can see, our use of the javap command simplified the printing of the decompiled bytecode. Now, let's use the -c parameter to reveal the bytecode:

```
$ javap -c CH1EX1.class
```

Here is the output from running the javap command with the -c parameter, using our example code:

```
Compiled from "CH1EX1.java"
public class CH1EX1 {
  public CH1EX1();
    Code:
       0: aload_0
       1: invokespecial #1                    // Method java/lang/
                                              Object."<init>":()V

       4: return

  public static void main(java.lang.String[]);
    Code:
       0: getstatic      #7                   // Field java/lang/System.
                                              out:Ljava/io/PrintStream;
       3: ldc            #13                  // String Basic Java loop.
       5: invokevirtual #15                   // Method java/io/
PrintStream.println:(Ljava/lang/String;)V
       8: iconst_1
       9: istore_1
      10: iload_1
      11: iconst_5
      12: if_icmpgt      34
      15: getstatic      #7                   // Field java/lang/System.
                                              out:Ljava/io/PrintStream;
      18: iload_1
      19: iload_1
      20: invokedynamic #21,  0               // InvokeDynamic
#0:makeConcatWithConstants:(II)Ljava/lang/String;
      25: invokevirtual #15                   // Method java/io/
PrintStream.println:(Ljava/lang/String;)V
      28: iinc           1, 1
      31: goto           10
      34: return
}
```

You should now have a better appreciation of the miraculous work the JVM does. We are just getting started!

## More options with javap

So far, we have used the javap command in two forms: without parameters and with the -c parameter. There are some other parameters we can use to further our understanding of our bytecode. To review these parameters, we will use an abbreviated example, shown next:

```
public class CH1EX2 {
    public static void main(String[] args) {
        System.out.println("Simple Example");
    }
}
```

You can now compile the class first using javac and then run it. Let's use the -sysinfo parameter to output system information. As you can see next, the file path, file size, date, and a SHA-256 hash are printed:

```
$ javap -sysinfo CH1EX2.class
```

Here is the output from using the -sysinfo parameter:

```
Classfile /HPWJ/Code/Chapter1/CH1Example2/src/CH1EX2.class
  Last modified Oct 22, 2023; size 420 bytes
  SHA-256 checksum
9328e8bab7fcd970f73fc9eec3a856a809b7a45e7b743f8c8b3b7ae0a7fbe0da
  Compiled from "CH1EX2.java"
public class CH1EX2 {
  public CH1EX2();
  public static void main(java.lang.String[]);
}
```

Let's look at one more example of the javap command, this time using the -verbose parameter. Here is how to use that parameter via a terminal command line:

```
$ javap -verbose CH1EX2.class
```

The output from using the -verbose parameter is shown next. As you can see, there is a lot of additional information provided:

```
Classfile /HPWJ/Code/Chapter1/CH1Example2/src/CH1EX2.class
  Last modified Oct 22, 2023; size 420 bytes
  SHA-256 checksum
9328e8bab7fcd970f73fc9eec3a856a809b7a45e7b743f8c8b3b7ae0a7fbe0da
  Compiled from "CH1EX2.java"
```

The remainder of the output can be obtained here: https://github.com/PacktPublishing/High-Performance-with-Java/tree/main/Chapter01/java-output.

We can use the javap command to thoroughly examine our bytecode. Here are some common parameters that are available to us, in addition to -c, -sysinfo, and -verbose, which we have already used:

| javap Parameter | What is Printed |
|---|---|
| -constants | Constants (static final) |
| -help (or -?) | Help |
| -l | Line variables and local variables |
| -private (or -p) | All classes |
| -protected | Protected and public classes (not private) |
| -public | Public classes (does not print private classes) |
| -s | Internal signature types |
| -version | Java release version |

Table 1.1 – javap parameters

Now that you have seen the inside workings of the JVM, let's take a look at the phenomenal job it does with garbage collection.

# Garbage collection

Java developers have long enjoyed the JVM's ability to manage memory, including allocation and deallocation. The allocation component of memory management is straightforward and not wrought with inherent problems. The area that is most important is freeing previously allocated memory that is no longer needed by the application. This is referred to as deallocation or garbage collection. While not unique to Java programming language, its JVM does a wonderful job with garbage collection. This section takes a detailed look at JVM's garbage collection.

## The garbage collection process

The following example is the creation of two objects, making them reference each other, and then nullifying them both. Once they are nullified, they are no longer reachable, despite that they reference each other. This makes them eligible for garbage collection:

```
public class Main {
    public static void main(String[] args) {
        // Creating two objects
        SampleClass object1 = new SampleClass("Object 1");
        SampleClass object2 = new SampleClass("Object 2");
```

```java
        // Making the objects reference each other
        object1.reference = object2;
        object2.reference = object1;

        // References are nullified
        object1 = null;
        object2 = null;


    }
}
class SampleClass {
    String name;
    SampleClass reference;

    public SampleClass(String name) {
        this.name = name;
    }

    // Overriding finalize() method to see the garbage collection
    // process
    @Override
    protected void finalize() throws Throwable {
        System.out.println(name + " is being garbage collected!");
        super.finalize();
    }
}
```

In the preceding example, we could make an explicit call to the garbage collector. This is not generally advisable as the JVM does a great job of this already, and extra calls could impact performance. If you do want to make that explicit call, this is how you would do it:

```java
System.gc();
```

> **Note on the finalize() method**
>
> The JVM only calls the `finalize()` method if it has been enabled. When enabled, it is possible that it will be called by the garbage collector, after an unspecified time delay. The method has been deprecated and should only be used for testing, not in production systems.

If you want to conduct additional testing, you can add code that attempts to reach unreachable objects. Here is how you would do that:

```java
try {
  object1.display();
```

```
  } catch (NullPointerException e) {
    System.out.println("Unreachable object!");
}
```

The preceding `try-catch` block has a call to `object1`. Since that is unreachable, it will throw a `NullPointerException` exception.

## Garbage collection algorithms

The JVM has several garbage collection algorithms at its disposal, and which one is used depends upon the version of Java and the specific use case. There are a few garbage collection algorithms worth mentioning:

- **Serial Collector**
- **Parallel Collector**
- **Concurrent Mark-Sweep Collector (CMS)**
- **Garbage-First (G1) Collector**
- **Z Garbage Collector (ZGC)**

Each of these garbage collection algorithms is described next so that you can gain a better appreciation for the heavy lifting the JVM does to deallocate memory for us.

### Serial Collector

The serial garbage collector is the JVM's most basic algorithm. It is used for small heaps and single-threaded applications. This type of application can be characterized as sequential execution, not concurrent execution. Since there is only one thread to consider, memory management is much easier. Another characterizing feature is that single-threaded applications often only use a portion of the host's CPU capabilities. Finally, since the execution is sequential, predicting the need for garbage collection is uncomplicated.

### Parallel Collector

The parallel garbage collector, also referred to as the throughput garbage collector, was the default for early Java releases through Java 8. Its capabilities exceed that of the serial collector in that it was designed for medium- to large-sized heaps and multithreaded applications.

### CMS

The CMS garbage collector was deprecated in Java 9 and removed from the Java platform in Java 14. Its purpose was to minimize the amount of time an application was paused for garbage collection. The reasons for deprecation and removal included high resource consumption, failures due to lag, difficulty in maintaining the code base, and that there were better alternatives.

**G1 Collector**

The G1 Collector became the default garbage collector in Java 9, dethroning the parallel collector. G1 was designed to provide faster and more predictable response times and extremely high throughput. The G1 Collector can be considered the primary and default garbage collector.

**ZGC**

Java 11 was released with the ZGC garbage collector as an experimental feature. ZGC's aim was to provide a low-latency garbage collector that was scalable. The intention was for ZGC to handle a wide range of heap sizes, from very small to very large (for example, terabytes). ZGC delivered on these goals without sacrificing pause times. This success led to ZGC being released as a Java 15 feature.

## Garbage collection optimizations

Discerning Java developers are laser-focused on getting the most performance out of the JVM for their applications. To this end, several things can be done with respect to garbage collection to help improve the performance of Java applications:

1. Select the most appropriate garbage collection algorithm based on your use case, heap size, and thread count.

2. Initialize, monitor, and manage your heap sizes. They should not be larger than absolutely needed.

3. Limit object creation. We will discuss alternatives to this in *Chapter 6*.

4. Use appropriate data structures for your applications. This is the subject of *Chapter 2*.

5. As indicated earlier in this chapter, avoid, or at least significantly limit, calls to the `finalize` method to reduce garbage collection delays and processing overhead.

6. Optimize your use of strings, with a special focus on minimizing duplicate strings. We will cover this in greater depth in *Chapter 7*.

7. Be mindful of allocating memory outside of the JVM's garbage collection reach (for example, native code).

8. Lastly, employ tools to help you monitor garbage collection in real time. We will review some of these tools in *Chapter 14*.

Hopefully, you have a newfound appreciation of JVM's garbage collection and optimization methods. We will now take a look at specific compiler optimizations.

# JIT compiler optimizations

There are three key components of the JVM: a class loader that initializes and links classes and interfaces; runtime data, which includes memory allocation; and the execution engine. This latter component, the execution engine, is the focus of this section.

The core responsibility of the execution engine is to convert bytecode so that it can be executed on the host **central processing unit (CPU)**. There are three primary implementations of this process:

- Interpretation
- **Ahead-of-Time (AOT)** compilation
- JIT compilation

We will take a cursory look at interpretation and AOT compilation before diving into JIT compilation optimizations.

## Interpretation

Interpretation is a technique that the JVM can use to read and execute bytecode without converting it into machine code native to the host CPU. We can invoke this mode by using the `java` command in a terminal window. Here is an example:

```
$ java Main.java
```

The only real advantage of using the interpretation mode is to save time by avoiding the compilation process. This can be useful for testing and is not recommended for production systems. Performance is a typical problem when using interpretation as compared to AOT and JIT, both described next.

## AOT compilation

We can compile our bytecode into native machine code in advance of application execution. This approach is called AOT and can be used for performance enhancements. Specific advantages to this approach include the following:

- You can avoid normal application startup delays associated with the traditional JIT compilation process
- The startup execution speed is consistent
- Startup CPU load is typically reduced
- Mitigate security risks by avoiding other compilation approaches
- You can build optimization into your startup code

There are also a few disadvantages to using the AOT compilation approach:

- When you compile ahead of time, you lose the ability to deploy on any device
- It is possible that your application will be bloated, increasing storage and related costs

- Your application will not be able to take advantage of JVM optimizations that are associated with JIT compilation
- The complexity of code maintenance increases

Understanding the benefits and disadvantages can help you decide when to, and when not to, use the AOT compilation process.

## JIT compilation

The JIT compilation process is likely the one we are most familiar with. We invoke the JVM and have it convert our bytecode into machine code specific to the current host machine. This machine code is referred to as native machine code because it is native to the local CPU. This compilation occurs just in time, or just before execution. This means the entirety of the bytecode is not compiled at once.

The advantages of JIT compilation are increased performance over the interpretation approach, the ability to deploy your application on any device (platform agnostic), and the ability to optimize. The JIT compiler process is capable of optimizing the code (for example, removing dead code, loop unrolling, and more. The disadvantages are the overhead required at the initial startup and the use of memory caused by the need to store native machine code translations.

## Summary

By now, you should have an appreciation of the complexities of the JVM and how it works. This chapter's coverage included the `javac` and `javap` command-line tools to create and analyze bytecode. The JVM's garbage collection function was also examined through the lens of application performance. Lastly, the optimization of JIT compilation was covered.

The JVM is 29 years old, as of this book's publication date, and it has come a long way since its initial release. In addition to continual improvements and optimizations, the JVM can even support additional languages (for example, Kotlin and Scala). Java developers interested in continual performance improvement of their Java applications should stay abreast of JVM updates.

With a solid understanding of the JVM, we turn our focus to data structures in the next chapter. Our focus will be on using data structures optimally as part of our high-performance strategy.

# 2

# Data Structures

Data structures are important components that contribute to, or detract from, the performance of our Java applications. They are foundational parts that are used throughout our programs and can help us organize and manipulate data efficiently. Data structures are essential for optimizing the performance of our Java applications because they can ensure that our data access, memory management, and caching are efficient. Proper use of data structures can lead to algorithm efficiency, the scalability of our solutions, and the safety of our threads.

The significance of data structures can be evidenced by reducing the time complexity of operations. With proper data structure implementation, we can improve the predictability and consistency of our application's performance. In addition to improving the performance of our Java applications, properly chosen data structures result in increased code readability, making them easier to maintain.

In this chapter, we are going to cover the following main topics:

- Lists
- Arrays
- Trees
- Stacks and queues
- Advanced data structures

By the end of this chapter, you should understand how specific data structures, such as **lists**, **arrays**, **trees**, **stacks**, and **queues**, can impact the performance of Java applications. You will have the opportunity to gain hands-on experience with Java code that demonstrates how to improve performance through proper data structure selection and implementation.

## Technical requirements

To follow the examples and instructions in this chapter, you will need to be able to load, edit, and run Java code. If you haven't set up your development environment, please refer to *Chapter 1*.

The code for this chapter can be found here: `https://github.com/PacktPublishing/High-Performance-with-Java/tree/main/Chapter02`.

# Improving performance with lists

A list is a fundamental data structure in the Java programming language. They give us the ability to easily create, store, and manipulate an ordered collection of elements. This data structure uses the `java.util.list` interface and extends the `java.util.Collection` interface.

In this section, we will take a close look at lists, why and when to use them, and techniques for getting the highest performance out of them.

## Why use lists?

Perhaps the most common way of explaining what the list data structure can be used for is as a check-off/to-do list or a grocery shopping list. We create lists in our programs because we want to leverage one or more of its advantages:

- **Ordered elements**: Lists are used to maintain the order of our elements, even as we add new elements. This allows us to manipulate our data in a specific sequence. Consider a system log that has new entries added with date and time stamps. We would want those entries to be maintained in a specific order.

- **Automatic resizing**: Lists can be used to dynamically resize themselves as our programs add and remove elements. This is especially true in **ArrayLists**, which are covered later in this chapter.

- **Positional data access**: Lists allow us to obtain random access efficiently by using the element's index, also referred to as **positional data**.

- **Duplicate elements**: Lists allow us to have duplicate elements. So, if this is important for your use case, then you might consider using a list as your data structure selection.

- **Iterability**: The list data structure allows us to easily iterate through our elements. We can use loops and the `forEach` method. We will look at an example of this in the next section of this chapter.

- **Multiple implementation options**: Java's list data structure can be implemented as a `LinkedList`, `ArrayList`, or `Vector` list type. These list types offer unique characteristics. The following table shows the different characteristics of these list types. Pay special attention to the performance row:

| | **LinkedList** | **ArrayList** | **Vector** |
|---|---|---|---|
| **Data structure** | Doubly-linked list | Dynamic array | Dynamic array |
| **Use case** | To frequently manipulate data | For fast reads | When thread safety is required |

| | **LinkedList** | **ArrayList** | **Vector** |
|---|---|---|---|
| **Performance** | + Adds and deletes<br><br>- Access by index | - Adds and deletes<br><br>+ Access by index | - Adds and deletes<br><br>- Access by index |
| **Thread safety** | No, not by default | No, not by default | Yes, by default |

Table 2.1 – Lists

When choosing between `LinkedList`, `ArrayList`, and `Vector` list types, we should consider the requirements based on our use case. For example, if our use case includes frequent adds and deletes, then `LinkedList` might be the best option. Alternatively, if we have infrequent adds and deletes, but heavy reads, then `ArrayList` is likely our best option. Finally, if we are most concerned about thread safety, `Vector` might be our best option.

> **Important note about thread safety**
>
> While `LinkedList` and `ArrayList` are not thread-safe by default, they can be made thread-safe by explicitly synchronizing access. This should prevent concurrent access-related data corruption, but it will likely lead to lower performance regarding your Java application.

- **Build-in methods**: Perhaps one of the greatest benefits of using a list is that we can take advantage of methods in the `java.util.List` interface, which the list implements. These built-in methods include functionality for searching, adding, and removing elements. A full list of `java.util.Llist` methods are available in the official Java documentation: `https://docs.oracle.com/javase/8/docs/api/java/util/List.html`.

Now that we've reviewed why we should use lists, let's look at examples of common implementations.

## Common list implementations

In this section, we will look at implementation examples for `ArrayList`, `LinkedList`, and `Vector`.

### ArrayList example

Our first list implementation example is an `ArrayList` list type of numbers. We will assume that this is part of a human resources (HR) system that stores start dates, end dates, and length of service. As shown in the following code, we must import both `java.util.ArrayList` and `java.util.List`:

```
import java.util.ArrayList;
import java.util.List;
```

Next, we have our class declaration and main method. Here, we must create an `ArrayList` list type called `hr_numbers`:

```
public class Example1 {
  public static void main(String[] args) {
    List<Integer> hr_numbers = new ArrayList<>();
```

The next two lines of code use the `.add` method to add elements to our `ArrayList`:

```
    hr_numbers.add(1983);
    hr_numbers.add(2008);
```

The next three lines of code get the first two numbers, using the `get` method, and use them to calculate a value to add as a third element to `ArrayList`:

```
    int startYear = hr_numbers.get(0);
    int endYear = hr_numbers.get(1);
    hr_numbers.add(endYear-startYear);
```

The last section of the code is a `for` loop. This iterates through the list and provides output to the terminal window:

```
    for (int number : hr_numbers) {
      System.out.println(number);
    }
```

Here's the program's output:

```
1983
2008
25
```

As you can see, our output is as expected; we simply print each of the three elements to the terminal window using a `for-each` loop.

> **for-each loops**
>
> Java 5 introduced an enhanced `for` loop called the `for-each` loop. We can use this loop to iterate through elements without using an explicit iterator or index. This makes our code quicker to write and more readable.

## LinkedList example

Our LinkedList implementation example is a simple **LinkedList** of pet names. The following code demonstrates how to create a list and use the get, remove, contains, and size methods. As you can see, we import both java.util.LinkedList and java.util.List:

```
Import java.util.LinkedList;
import java.util.List;
```

Next, we have our class declaration and the main method. Here, we create a LinkedList list type called petNames:

```
public class Example2 {
  public static void main(String[] args) {
    List<String> petNames = new LinkedList<>();
```

The next four lines of code use the .add method to add elements to our LinkedList:

```
petNames.add("Brandy");
petNames.add("Muzz");
petNames.add("Java");
petNames.add("Bougie");
```

The next two lines of code demonstrate how to get the first two pets using the get method:

```
String firstPet = petNames.get(0);
String secondPet = petNames.get(1);
```

The next section of the code is a for-each loop that iterates through the list and provides output to the terminal window:

```
for (String pet : petNames) {
  System.out.println(pet);
}
```

Here's the loop's output:

```
Brandy
Muzz
Java
Bougie
```

We can remove an element from our LinkedList by using the remove method, as illustrated here. As you can see, after calling the remove method, Brandy is no longer an element in LinkedList:

```
petNames.remove("Brandy");
```

The next part of our code calls the `contains` method to check if a specific value is found in `LinkedList`. Since we previously removed this pet from `LinkedList`, the Boolean result is `false`:

```
boolean containsBrandy = petNames.contains("Brandy");
System.out.println(containsBrandy);
```

The output of the `println` statement is as expected:

```
false
```

This final section of our code demonstrates the use of the `size` method. Here, we make a call to that method, which returns an integer. We use that value in our final output:

```
int size = petNames.size();
System.out.println("You have " + size + " pets.");
```

The final output reflects the expected size of our `LinkedList`:

```
You have 3 pets.
```

Now that we've looked at how to implement `ArrayList` and `LinkedList`, let's look at our final example, **Vector**.

## Vector example

Our last list implementation example is a `Vector` list type. As you will see, `Vector` is similar to `ArrayList`. We implement them as dynamic arrays so that we can benefit from efficient random access to our vector's elements. Vectors have the added benefit of being thread-safe by default. This is due to the default synchronization we previously discussed. Let's look at some example code.

Our example program will store a set of lucky numbers. It starts by importing the `java.util.Vector` and `java.util.Enumeration` packages:

```
import java.util.Vector;
import java.util.Enumeration;
```

Next, we have our class declaration and the main method. Here, we create a `Vector` list type called `luckyNumbers` that will store integers:

```
public class Example3 {
  public static void main(String[] args) {
    Vector<Integer> luckyNumbers = new Vector<>();
```

The next three lines of code use the .add method to add elements to our Vector.

```
luckyNumbers.add(8);
luckyNumbers.add(19);
luckyNumbers.add(24);
```

The next two lines of code demonstrate how to access a vector's elements with the use of the index and the get method. Note that indices start at zero (0):

```
int firstNumber = luckyNumbers.get(0);
int secondNumber = luckyNumbers.get(2);
```

The next section of the code uses a legacy approach for iterating through a Vector list type. The enumeration approach can be used in place of an enhanced or for-each loop. Vectors, in and of themselves, are considered a list type that's falling out of common use. The following code iterates through the list and provides output to the terminal window:

```
Enumeration<Integer> enumeration = luckyNumbers.elements();
while (enumeration.hasMoreElements()) {
    int number = enumeration.nextElement();
    System.out.println(number);
}
```

Here's the program's output:

```
8
19
24
```

We can remove an element from our Vector by using the removeElement method, as illustrated here. After calling the removeElement method, the lucky number 19 is removed as an element from Vector:

```
luckyNumbers.removeElement(19);
```

The next part of our code calls the contains method to check if a specific value is found in Vector. Since we previously removed this lucky number from Vector, the Boolean result is false:

```
boolean containsNineteen= luckyNumbers.contains(19);
System.out.println(containsNineteen);
```

The output of the println statement is as expected:

```
false
```

This final section of our code demonstrates the use of the `size` method. Here, we make a call to that method, which returns an integer. We use that value in our final output:

```
int mySize = luckyNumbers.size();
System.out.println("You have " + mySize + " lucky numbers.");
```

The final output reflects the expected size of our `LinkedList` list type:

```
You have 2 lucky numbers.
```

This section provided examples of using lists for `ArrayList`, `LinkedList`, and `Vector`. There are additional implementations that you can consider, including `CopyOnWriteArrayList`, `CopyOnWriteArraySet`, and `LinkedHashSet`. Stacks are another implementation and are covered later in this chapter. Next, we will look at achieving high performance with lists.

## High performance with lists

Let's take a look at how we can improve the performance of our Java apps when using lists. The following code example implements a `LinkedList` list type of integers. Here, we'll create a `LinkedList` list type, add four elements to it, and then iterate through the list, printing each element to the screen:

```
import java.util.LinkedList;
import java.util.List;
public class Example4 {
  public static void main(String[] args) {
    List<Integer> numbers = new LinkedList<>();
    numbers.add(3);
    numbers.add(1);
    numbers.add(8);
    numbers.add(9);
    System.out.println("Initial LinkedList elements:");
    for (int number : numbers) {
      System.out.println(number);
    }
  }
}
```

The output of this first section is as follows:

```
Initial LinkedList elements:
3
1
8
9
```

The next section of code removes an element using the `remove` method – specifically, the first occurrence of 8:

```
numbers.remove(Integer.valueOf(8));
```

The following code segment performs two checks using the `contains` method. First, it checks for a 3 and then an 8. The results are printed on the screen:

```
boolean containsThree = numbers.contains(3);
System.out.println("\nThe question of 3: " + containsThree);
boolean containsEight = numbers.contains(8);
System.out.println("The question of 8: " + containsEight);
```

The output of this section of code is shown here:

```
The question of 3: true
The question of 8: false
```

This final segment of code iterates through `LinkedList` and prints its values:

```
System.out.println("\nModified LinkedList elements:");
for (int number : numbers) {
   System.out.println(number);
}
```

The output of this final section of code is as follows:

```
Modified LinkedList elements:
3
1
9
```

Now, it's time to see what we can do to modify our code to improve overall performance. There are several techniques we can use:

- **Data type usage**: We should always use the most appropriate data types for our LinkedLists. In our example, we used `List<Integer>` objects. The `Integer` class is essentially a wrapper around the primitive `int` data type. The `Integer` object only contains one field of the `int` type. Here's how we can modify our code:

  ```
  LinkedList<Integer> numbers = new LinkedList<>();
  ```

- **Element removal**: When we remove elements from LinkedList, we should use an **iterator**. To use an iterator, we need to import the java.util.Iterator package. This is an efficient method of avoiding errors such as **ConcurrentModificationsException**, which is thrown when we attempt to make two simultaneous modifications to a collection. Here's how we can write such an iterator:

```
Iterator<Integer> iterator = numbers.iterator();
while (iterator.hasNext()) {
  int number = iterator.next();
  if (number == 8) {
    iterator.remove();
  }
}
```

The fully revised example is available at https://github.com/PacktPublishing/High-Performance-with-Java/tree/main/Chapter02/Example5.java.

You should now have increased knowledge and confidence in using lists in Java to help improve the performance of your applications. Next, we will look at arrays and how to use them optimally.

## Improving performance with arrays

Previously, we discussed ArrayList; in this section, we will focus on arrays. There are key differences in these two data types: arrays have a fixed size, while ArrayLists do not. In this section, we will review the characteristics of arrays and how to improve the performance of our Java applications when implementing arrays.

### Array characteristics

There are four primary characteristics of arrays. Let's take a look:

- **Size**: The size of an array is determined when it's created. It cannot be changed when the application is running. This fixed-sized characteristic is also referred to as **static**. Here's the syntax for creating an array:

```
int[] myNumbers = new int[10];
```

As you can see, we explicitly specify a size of 10 as part of the array's declaration.

- **Homeogeneous**: Homeogeneous means that all data in an array must be of the same kind. As with the previous example, we created an array of integers. We could have used strings, but we cannot mix data types in a single array. We can also have an array of objects and arrays of arrays.

- **Indexing**: This should go without saying, so as a reminder, our indexes start with 0, not 1. So, the array that we previously created has 10 elements, indexed from 0 to 9.

- **Continuous memory**: One of the great efficiencies offered to us by arrays is random access. This efficiency stems from the fact that all elements in an array are stored in contiguous memory. In other words, the elements are stored in memory locations next to each other.

Now that we have a firm understanding of array characteristics, let's explore some code that implements this important data structure.

## Implementing arrays

This section features a basic Java application that implements an array of planets using the `String` data type. As we walk through the code, we will create the array, access and print array elements, use the length method, access an array element using its index, and modify an array.

This first section of code creates an array of **strings**:

```
public class Example6 {
    public static void main(String[] args) {
        String[] planets = {
            "Mercury",
            "Venus",
            "Earth",
            "Mars",
            "Jupiter",
            "Saturn",
            "Uranus",
            "Neptune"
        };
    }
}
```

As you can see, we created an array of eight strings. This next section of code demonstrates accessing and printing all elements of an array:

```
System.out.println("Planets in our solar system:");
for (int i = 0; i < planets.length; i++) {
    System.out.println(planets[i]);
}
```

Here's the output of the preceding code snippet:

```
Planets in our solar system:
Mercury
Venus
Earth
Mars
Jupiter
Saturn
Uranus
Neptune
```

We can use the `length` method to determine the size of our array. It might seem like we don't need to determine the size since it is determined when the array is created. Often, we don't know the initial size of our arrays because they're based on external data sources. Here's how to determine an array's size:

```
int numberOfPlanets = planets.length;
System.out.println("Number of planets: " + numberOfPlanets);
```

The output of the preceding code is shown here:

```
Number of planets: 8
```

Next, we will look at how to access an array element by referencing its index position within the array:

```
String thirdPlanet = planets[2];
System.out.println("The third planet is: " + thirdPlanet);
```

The output of the previous two lines of code is as follows:

```
The third planet is: Earth
```

As you can see, we printed the third element in our array by using an index reference of 2. Remember, our indexes start with 0.

Our last segment of code shows how we can modify an element in our array:

```
planets[1] = "Shrouded Venus";
System.out.println("After renaming Venus:");
for (String planet : planets) {
  System.out.println(planet);
}
```

The output of the preceding code is shown here. As you can see, the new name for the element at index position 1 is reflected in the output:

```
After renaming Venus:
Mercury
Shrouded Venus
Earth
Mars
Jupiter
Saturn
Uranus
Neptune
```

The fully revised example is available at https://github.com/PacktPublishing/High-Performance-with-Java/tree/main/Chapter02/Example6.java.

Now that you've learned how to implement arrays in Java, let's look at some approaches to improving our application's performance in terms of array usage.

## High performance with arrays

As we have seen with lists, our implementation approaches to data structures can have a direct impact on our Java application's performance. This section documents several strategies and best practices for optimizing array-related Java code. These strategies can be categorized as algorithmic optimization, data structures, memory management, parallel processing, vectorization, caching, and benchmarking and profiling. Let's look at each of them.

### Algorithmic optimization

Whenever we select algorithms, we should ensure they are the most appropriate for the data types we're using. For example, it's important to select the most efficient sorting algorithm (that is mergesort, quicksort, and so on) based on our array's size and characteristics. In addition, binary searches can be implemented when searching in sorted arrays.

Additional information is provided in *Chapter 5* of this book.

### Data structures

The core of this chapter focuses on selecting the right data structure based on your use case and requirements. When it comes to arrays, we should select them when we need frequent read access; here, our dataset can be of a fixed size. In addition, knowing when to select `ArrayList` over `LinkedList` is equally important.

## Memory management

Whenever possible, we should avoid creating temporary objects to support array operations. Instead, we should reuse arrays or use methods such as `Arrays.copyOf` or `System.arraycopy` for greater efficiency.

Additional information is provided in *Chapter 8* of this book.

## Parallel processing

When we must sort and process large arrays, it can be beneficial to employ Java's parallel processing capabilities, such as by using `parallelSort`. You should consider using multithreading with concurrent array processing. This is especially important for large arrays.

Additional information is provided in *Chapter 9* of this book.

## Vectorization

In the context of Java arrays, vectorization is a technique that involves performing operations on more than one element of an array simultaneously. This typically includes optimizing modern central processors. The goal is to increase array processing operations. This is often referred to as **single instruction, multiple data** (SIMD). There are specific data types that are designed to work with SIMD instructions so that they include the `java.util.Vector` class, which was introduced with Java 16.

Vectorization can provide significant performance benefits, especially with large arrays. This is exponentially true when operations can be parallelized. There are limits to what can be vectorized, such as dependencies and complex operations.

## Caching

A proven performance approach is to optimize array access patterns to support cache locality. This can be achieved by accessing contiguous memory locations. Another approach is to minimize pointer aliasing to the best extent possible. While point aliasing might support compiler optimizations, local variables or array indexes should be used instead for optimal performance.

Additional information is provided in *Chapter 8* of this book.

## Benchmarking and profiling

Whenever possible, we should conduct benchmarking so that we can compare our approach to array operations. Armed with this analysis, we can select the most efficient and proven approach. As part of our analysis, we can employ profiling tools to help identify performance bottlenecks specific to our array operations.

Additional information is provided in *Chapter 13* of this book.

Now that you've learned how to improve Java application performance when dealing with arrays, let's look at how this can be done with trees.

# Improving performance with trees

A **tree** is a hierarchical data structure that consists of parent and child **nodes**, analogous to that of a physical tree. The topmost node is referred to as the **root** and there can only be one in a tree data structure. The nodes are the foundational components of this data structure. Each node contains both data and references to child nodes. There are additional terms that you should be familiar with regarding trees. A **leaf node** has no child. Any node and its descendants can be considered a **subtree**.

## Examples of a tree structure

As shown in the following example, we can implement trees as objects or classes:

```
class TreeNode {
   int data;
   TreeNode left;
   TreeNode right;
   public TreeNode(int data) {
     this.data = data;
     this.left = null;
     this.right = null;
   }
}
```

The preceding code snippet defines a TreeNode class that can be used along with another class for tree operation management. We need to create our tree in our main() method. Here's an example of how to do that:

```
Example7BinarySearchTree bst = new Example7BinarySearchTree();
```

To insert data into our tree, we can use the following code:

```
bst.insert(50);
bst.insert(30);
bst.insert(70);
bst.insert(20);
bst.insert(40);
bst.insert(60);
bst.insert(80);
```

To use or search our tree, we can perform a binary search for an element in our tree using code similar to the following:

```
int searchElement = 70;
if (bst.search(searchElement)) {
  System.out.println("\n" + searchElement + " was found in the tree.");
} else {
  System.out.println("\n" + searchElement + " was not found in the
  tree.");
}
```

A full working example is available at https://github.com/PacktPublishing/High-Performance-with-Java/tree/main/Chapter02/Example7.java.

## High-performance considerations

The tree data structure can be very complex, so we should consider our Java application's overall performance, especially regarding algorithms that manipulate our trees. Additional considerations can be categorized as type, safety, iteration, memory, operations, and caching. Let's briefly look at each of these categories.

### Type

It's important to carefully construct our trees so that they are as balanced as possible. We should be mindful of the **tree's height**, which is the longest path from the root to a leaf node. The data requirements should be reviewed so that we can select the most optimal tree. For example, we can use a **binary search tree**, which can provide us with efficient searches.

### Safety

As with other data structures, you should always implement thread safety protocols or, at a minimum, use concurrent tree structures.

### Iteration

Iteration operations can often present bottlenecks, calling for an optimization strategy. In addition, the use of recursive algorithms should be minimized to improve overall performance.

### Memory

Generally, and as it applies to trees specifically, we should minimize object creation and overhead. We should strive to do our best to manage memory efficiently.

Additional information is provided in *Chapter 8* of this book.

## Operations

There are several things we can do to optimize operations. For example, we can use batch operations to reduce processing overhead. Optimizing our algorithms can help us avoid excessive processing and data updates.

Additional information is provided in *Chapter 5* of this book.

## Caching

We can optimize the way our data is laid out to maximize cache locality. Our goal with caching is to reduce memory access times. In addition, we can access nearly all nodes in memory to further improve performance.

Additional information is provided in *Chapter 8* of this book.

# Improving performance with stacks and queues

**Stacks** and **queues** are almost always grouped because they can both be used to manage and manipulate collections of data elements. They are both linear stacks; that's where their similarities end. While grouped, there are key differences in how they operate. In this section, we'll look at how to implement stacks and queues and how to optimize them for high-performance Java applications.

## Implementing stacks

Stacks are linear data structures that use the **last in, first out** (**LIFO**) principle of element management. With stacks, elements are added to and removed from the **top** of the stack. We push elements to the top, peek to view an element, and pop to remove the top element.

The following example demonstrates how to create a stack in Java. You will notice that we start by importing the java.util.Stack package:

```
import java.util.Stack;
public class Example8 {
  public static void main(String[] args) {
    Stack<Double> transactionStack = new Stack<>();
    transactionStack.push(100.0);
    transactionStack.push(-50.0);
    transactionStack.push(200.0);
    while (!transactionStack.isEmpty()) {
      double transactionAmount = transactionStack.pop();
      System.out.println("Transaction: " + transactionAmount);
    }
  }
}
```

The preceding code is a simple implementation of a Java stack. The purpose of the program is to process bank transactions. Now that you've seen this simple implementation, let's see how we can refine our code for greater performance.

## Improving the performance of stacks

Using our previous example as a starting point, we will refine it for higher performance at runtime. We'll start by using a custom stack implementation. This type of refinement is especially effective when use cases involve a high volume of transactions. As you can see, we must import the `java.util.EmptyStackException` package.

Next, we must declare a class and use an array with a **double** data type. We're opting for this approach to avoid processing overhead due to **auto-boxing**:

```
public class Example9 {
  private double[] stack;
  private int top;

  public Example9(int capacity) {
    stack = new double[capacity];
    top = -1;
  }
}
```

The following segment of code establishes a method for push, pop, and isEmpty:

```
public void push(double transactionAmount) {
  if (top == stack.length - 1) {
    throw new RuntimeException("Stack is full.");
  }
  stack[++top] = transactionAmount;
}
public double pop() {
  if (isEmpty()) {
    throw new EmptyStackException();
  }
  return stack[top--];
}
public boolean isEmpty() {
  return top == -1;
}
```

Our last section of code is our `main()` method, which processes the stack:

```java
public static void main(String[] args) {
    Example9 transactionStack = new Example9(10);

    transactionStack.push(100.0);
    transactionStack.push(-50.0);
    transactionStack.push(200.0);

    while (!transactionStack.isEmpty()) {
        double transactionAmount = transactionStack.pop();
        System.out.println("Transaction: " + transactionAmount);
    }
}
```

The output for both sets of programs, simple and optimized, are the same and are shown here:

```
Transaction: 200.0
Transaction: -50.0
Transaction: 100.0
```

Full working examples of both versions of our stack implementations are available at `https://github.com/PacktPublishing/High-Performance-with-Java/tree/main/Chapter02/Example8.java` and `https://github.com/PacktPublishing/High-Performance-with-Java/tree/main/Chapter02/Example9.java`.

## Implementing queues

A queue is another linear data structure and uses the **first in, first out (FIFO)** principle. With queues, we add elements to the **rear** and remove them from the **front**. We can use a method such as enqueue, peek, and dequeue to manage our queues.

The following example demonstrates how to create a queue in Java. As you can see, we start by importing the `java.util.LinkedList` and `java.util.Queue` packages:

```java
import java.util.LinkedList;
import java.util.Queue;

public class Example10 {
    public static void main(String[] args) {
        Queue<Double> transactionQueue = new LinkedList<>();
        transactionQueue.offer(100.0);
        transactionQueue.offer(-50.0);
        transactionQueue.offer(200.0);
```

```
    while (!transactionQueue.isEmpty()) {
        double transactionAmount = transactionQueue.poll();
        System.out.println("Transaction: " + transactionAmount);
    }
    }
}
```

The preceding code is a simple implementation of a Java queue. The purpose of the program, just like that of the stack example, is to process bank transactions. Now that you've seen this simple implementation, let's see how we can refine our code for greater performance.

## Improving the performance of queues

Our refined version of a queue implementation has been optimized for runtime performance. This is especially important for high transactional applications. We start by importing the java.util. NoSuchElementException package, after which we declare the class and a set of private class variables before creating a constructor for our custom queue implementation:

```java
import java.util.NoSuchElementException;
public class Example11 {
    private double[] queue;
    private int front;
    private int rear;
    private int size;
    private int capacity;
    public Example11(int capacity) {
        this.capacity = capacity;
        queue = new double[capacity];
        front = 0;
        rear = -1;
        size = 0;
    }
```

The following section of code includes methods for enqueue, dequeue, and isEmpty:

```java
public void enqueue(double transactionAmount) {
    if (size == capacity) {
        throw new RuntimeException("Queue is full.");
    }
    rear = (rear + 1) % capacity;
    queue[rear] = transactionAmount;
    size++;
}
```

```java
public double dequeue() {
  if (isEmpty()) {
    throw new NoSuchElementException("Queue is empty.");
  }
  double transactionAmount = queue[front];
  front = (front + 1) % capacity;
  size--;
  return transactionAmount;
}

public boolean isEmpty() {
  return size == 0;
}
```

Our last section of code is our main() method, which processes the queue:

```java
public static void main(String[] args) {
  Example11 transactionQueue = new Example11(10);
  transactionQueue.enqueue(100.0);
  transactionQueue.enqueue(-50.0);
  transactionQueue.enqueue(200.0);

  while (!transactionQueue.isEmpty()) {
    double transactionAmount = transactionQueue.dequeue();
    System.out.println("Transaction: " + transactionAmount);
  }
}
```

The output for both sets of programs, simple and optimized, are the same and are provided here:

```
Transaction: 100.0
Transaction: -50.0
Transaction: 200.0
```

Full working examples of both versions of our queue implementations are available at https://github.com/PacktPublishing/High-Performance-with-Java/tree/main/Chapter02/Example10.java and https://github.com/PacktPublishing/High-Performance-with-Java/tree/main/Chapter02/Example11.java.

Both stacks and queues can be used for multiple use cases in Java. It's important to ensure we use them when it makes the most sense from a performance standpoint. Furthermore, we must consider the previously detailed optimization approaches.

# Improving performance with advanced data structures

There are more data structures available to us than the ones covered in this chapter thus far. Sometimes, instead of improving our use of a data structure, such as a list, array, tree, stack, or queue, we should implement more advanced data structures that are tailored to our specific use case and then optimize their use.

Let's look at some additional, more advanced, data structures that have performance considerations.

## Hash tables

We can use **hash tables** when we need fast **key-value** lookups. There are multiple hash functions to choose from and you should be mindful of hash collisions – they need to be managed efficiently. Some additional considerations are load factor, resizing, memory usage, and performance.

Here's an example of how to create a hash table:

```
import java.util.HashMap;
HashMap<String, Integer> hashMap = new HashMap<>();
hashMap.put("Alice", 25);
```

## Graphs

Graphs, such as **matrices** or **adjacency lists**, can be used to network an application and model complex data relationships. When implementing graphs, we should implement algorithms that transverse the graph efficiently. Two algorithms that are worth exploring are **breadth-first search** and **depth-first search**. Comparing the efficiencies of these algorithms can help you select the most optimized solution.

Here's how we can implement a graph:

```
import java.util.ArrayList;
import java.util.List;
List<List<Integer>> graph = new ArrayList<>();
int numNodes = 5;
for (int i = 0; i < numNodes; i++) {
  graph.add(new ArrayList<>());
}
graph.get(0).add(1);
```

## Trie

A **trie** is a data structure that is useful when you're programming methods for auto-completion, prefix matching, and more. Tries allow us to efficiently store and search for string sequences.

Here's an example of how to create a trie and create a root node for it:

```
class TrieNode {
  TrieNode[] children = new TrieNode[26];
    boolean isEndOfWord;
}
TrieNode root = new TrieNode();
```

We will cover string manipulation in *Chapter 7* of this book.

## Heap

**Priority queues** are usually implemented as **heaps** because they can efficiently manage elements based on their order or priority. This can be an ideal data structure for use cases that include scheduling, prioritizing, and element order.

Here's an example of how to implement a priority queue as a heap. This example demonstrates both minHeap and maxHeap and adds an element to each:

```
import java.util.PriorityQueue;

PriorityQueue<Integer> axheap = new PriorityQueue<>();
axheap.offer(3);

PriorityQueue<Integer> axheap = new PriorityQueue<>((a, b) -> b - a);
axheap.offer(3);
```

## Quad trees

A **quad tree** is a two-dimensional partition that's used to organize spatial data efficiently. They can be useful for spatial indexing, geographic information systems, collision detection, and more.

Creating a quad tree is a bit more involved than implementing other data structures. Here's a straightforward approach:

```
class QuadTreeNode {
  int val;
  boolean isLeaf;
  QuadTreeNode topLeft;
  QuadTreeNode topRight;
  QuadTreeNode bottomLeft;
  QuadTreeNode bottomRight;

  public QuadTreeNode() {}
  public QuadTreeNode(int val, boolean isLeaf) {
```

```
        this.val = val;
        this.isLeaf = isLeaf;
    }
}
QuadTreeNode root = new QuadTreeNode(0, false);
```

## Octrees

An **octree** is a three-dimensional partition that's used to organize spatial data efficiently. Like quad trees, octrees can be useful for spatial indexing, geographic information systems, collision detection, and more.

Octrees can be difficult to implement in Java, as is true with most three- or greater-dimensional data structures.

## Bitsets

When we need space-efficient storage for integer or Boolean datasets and also want to ensure their manipulation is efficient, the **bitset** data structure is worth considering. Example use cases include algorithms that traverse graphs and mark visited nodes.

Here's an example of how to implement a bitset:

```
import java.util.BitSet;
BitSet bitSet = new BitSet(10); // Creates a BitSet with 10 bits
bitSet.set(2);
boolean isSet = bitSet.get(2);
```

## Ropes

Ropes are great for implementing efficient string handling, especially large strings. They are used for performing concatenation and substring operations. Some example use cases are text editing applications and string manipulation functionality.

Ropes are extremely complex data structures to implement from scratch. Developers often turn to packages and tools that are external to the JDK to do this.

We will cover string manipulation in *Chapter 7* of this book.

These are just some of the advanced data structures available to us in Java. When incorporating these structures, we must understand how they work, their strengths, and their weaknesses.

# Summary

This chapter took an in-depth look at several data structures and demonstrated their significance to a Java application's overall performance. We embraced the essential nature of data structures for optimizing the performance of our Java applications because they can ensure that our data access, memory management, and caching are efficient. Proper use of data structures can lead to algorithm efficiency, the scalability of our solutions, and the safety of our threads.

Through code examples, we looked at nonoptimized and optimized examples of code for lists, arrays, trees, stacks, and queues. We also explored several advanced data structures and examined their added complexities. When we implement data structures properly, we can improve the predictability and consistency of our application's performance and make our code more readable and easier to maintain.

In the next chapter, we will focus on loops and how to optimize them for increased performance regarding our Java applications. This is a natural progression following our coverage of data structures as we will use them in our examples in the next chapter.

# 3

# Optimizing Loops

Loops are fundamental programming constructs that are not terribly difficult to understand or write. We use them to iterate through our application's data structures and perform repetitive tasks. We often take loops for granted based on their simple syntax and readability. When performance is a concern, loops have a duality. On one side, loops serve as a fundamental construct for efficient data processing. On the other side, poorly optimized loops can introduce significant bottlenecks and degrade the overall performance of our Java applications.

Concepts covered in this chapter include loop overhead, loop unrolling, benchmarks, loop fusion, loop parallelization, and loop vectorization. We will use code examples to provide insights and demonstrate best practices.

This chapter covers the following main topics:

- Types of loops
- Testing loops for performance
- Nested loops

This chapter explores techniques, strategies, and best practices to help you get the best performance out of your loops and to prevent the unintentional introduction of substantial bottlenecks that undermine runtime performance.

## Technical requirements

To follow the examples and instructions in this chapter, you will need the ability to load, edit, and run Java code. If you have not set up your development environment, please refer back to *Chapter 1*.

The finished code for this chapter can be found here: https://github.com/PacktPublishing/High-Performance-with-Java/tree/main/Chapter03

# Types of loops

Loops are indispensable constructs for iterating through data structures, controlling the flow of code, and performing repetitive tasks. They are central to many algorithms and applications. While loops can be used to efficiently process data and perform repetitive tasks, they can be equally problematic when we are concerned with performance.

It is important to understand different types of loops, their characteristics, and their performance implications. This section explores different loop types in Java, which include for, while, do-while, and for-each. Our goal is to understand each loop type's purpose and impact on code readability and efficiency. Specifically, we will cover the following topics:

- A loop's impact on performance
- Loop optimization basics
- Loop unrolling

Being armed with a deeper understanding of loop types can equip you to make the right loop selection for a given requirement.

## A loop's impact on performance

In order to understand a loop's impact on performance, we must have a firm understanding of the different types of loops. This section provides information on each loop type to include use cases and advantages. Examples are used to provide implementation details.

### *for loops*

for loops are the most basic loop type. They have a concise syntax with well-defined loop control. They are the unofficial default for iterating through a set of items. Let's look at the following components of these loops: initialization, condition, iteration, and efficiency.

The syntax for a for loop is as follows:

```
for (int i = 0; i < 5; i++); {}
{
```

Dissecting the preceding syntax, we can identify int  i  =0; as the **initialization** component of the loop. The **loop control variable** is i and it is set to 0. The i  <  5; **condition** checks if i is less than 5. If the evaluation returns true, then the loop continues; otherwise, it ends. The final component is the **iteration** where we use i++ to increase the **control variable** by 1 at the end of each iteration.

There are some performance-related issues with a `for` loop. When we initialize our loop, we should avoid initializing variables outside of the loop as it can sometimes degrade performance with redundant initializations within the loop. The **condition expression** directly impacts the number of iterations the loop will perform. Making this expression overly complex can slow down the loop. Lastly, the **iteration expression** should be properly designed to help prevent an infinite loop and inefficiency.

Let's look at an inefficient example and then a second example with performance improvements. The following example counts how many corgis have names with the letter "e":

```
Corgi[] corgis = getCorgiArray();
int count = 0;
for (int i = 0; i < corgis.length; i++) {
  if (corgis[i].getName().contains("e")) {
    count++;
  }
}
```

The preceding code has two inefficiencies:

* The loop condition is checked using the `length()` method in each iteration. This means that, for each corgi in the array, the length of the array is accessed, which can lead to unnecessary performance overhead.

* Our code snippet uses `corgis[i].getName().contains("e")` to check if the corgi's name contains the letter "e." So, for each corgi name, we are creating a new string and performing a string search. This will most assuredly be computationally expensive, especially with long names.

With these inefficiencies in mind, let's look at a revised section of that code:

```
Corgi[] corgis = getCorgiArray();
int count = 0;
int corgisLength = corgis.length;
for (int i = 0; i < corgisLength; i++) {
  String name = corgis[i].getName();
  if (name.indexOf('e') != -1) {
    count++;
  }
}
```

This code snippet is an improved version that has the following performance enhancements:

* We cached the array's length by storing it in the `corgislength` variable. Since we populate that variable outside of the loop, we avoid repeated access to the array's length during each iteration. This can save tremendously on computational effort.

- Instead of using the `contains()` method to check the names, we use the `indexof()` method. There is a significant difference here. When we use the `contains()` method, a new substring is created and a full search is conducted. Using the `indexof()` method returns the index position of the first occurrence of "e" or -1 if it is not found. This is a much more efficient way to check for a character's existence in a string without creating unnecessary substrings.

With our exploration of `for` loops completed, let us look at how to use `while` loops in an efficient manner.

## while loops

`while` loops are perhaps the second most common loop used in Java. They are fundamental flow control constructs that we can use to repeatedly execute a block of code, as long as a specified condition remains `true`.

The syntax for a `while` loop is as follows:

```
int count = 0;
int number = 1;
while (number <= 100) {
    count += number;
    number++;
}
```

These loops do not require initialization, but when initialization is employed, care should be taken to ensure they are correct so as to prevent unintended application behavior such as an infinite loop. When the **condition** evaluates as `true`, the body of the loop is executed. The loop terminates when the **condition** evaluates as `false`.

> **Condition expressions**
>
> We should strive to create straightforward condition expressions for loop efficiency. Overly complex condition expressions can increase processing overhead, resulting in poor performance.

Let's look at an inefficient example and then a second example with performance improvements. The following example is of an online ordering system. We want to check each order ID to determine if it starts with the "OL" prefix. If it does not, it should be added:

```
List<String> orderIDs = getOrderIDs();
int index = 0;
while (index < orderIDs.size()) {
    String orderID = orderIDs.get(index);
    if (!orderID.startsWith("OL")) {
        orderID = "OL" + orderID;
        orderIDs.set(index, orderID);
    }
```

```
    index++;
}
```

There are two inefficiencies in the preceding code snippet. First, the list size is repeatedly accessed; it is checked during each iteration. This can lead to unnecessary performance overhead. Secondly, the string concatenation, using the + operator, creates a new string, which can be inefficient when repeatedly performed within a loop.

The following code snippet is a modified version of the online ordering system:

```
List<String> orderIDs = getOrderIDs();
int index = 0;
int orderCount = orderIDs.size();
while (index < orderCount) {
    String orderID = orderIDs.get(index);
    if (!orderID.startsWith("OL")) {
        orderIDs.set(index, "OL" + orderID);
    }
    index++;
}
```

The preceding updated code snippet stands to improve system performance due to two factors. First, the list size is now cached, and second, we are using a more efficient string modification approach. Here, we are modifying the strings without creating new string objects.

With our exploration of while loops completed, let us look at how to use do-while loops in an efficient manner.

### do-while loops

do-while loops, as with the ones previously covered, are flow control structures that allow for the repeated execution of a code block while a specified condition remains true. The unique nature of this type of loop is that the condition is checked after the loop's code block is executed. This ensures that the loop's block of code is executed at least once.

The following example shows a simple do-while loop for a guessing game where the user continues to guess a number until they guess correctly:

```
int secretNumber = generateSecretNumber();
int userGuess;
boolean correctGuess = false;
do {
    userGuess = getUserGuess();
    if (userGuess == secretNumber) {
        correctGuess = true;
```

```
    } else {
        System.out.println("Try again!");
    }
} while (!correctGuess);
System.out.println("You guessed the secret number: " +
secretNumber);
```

As you can see in the preceding code snippet, we use a Boolean flag to control our loop. The flag is checked during each iteration. This approach is inefficient in that it adds unnecessary complexity and an extra variable. Let's modify the code so that it is more efficient:

```
import java.util.random;
int secretNumber = generateSecretNumber();
int userGuess;
    do {
        userGuess = getUserGuess();
        if (userGuess != secretNumber) {
            System.out.println("Try again!");
        }
    } while (userGuess != secretNumber);
    System.out.println("You guessed the secret number: " +
secretNumber);
```

With our updated example, we use userGuess != secretNumber as the loop condition, negating the need for a Boolean flag. This both simplifies our code and can make our code more efficient.

### for-each loops

We introduced the for-each loop in *Chapter 2* and noted that this type of loop is also referred to as an enhanced for loop. We choose to implement this type of loop when we want to iterate over all elements in a dataset, without having to manage an iterator or index. Let's look at an initial example and then refine it for better performance.

The following code snippet is an automotive parts processing application:

```
List<String> autoParts = getAutoParts();
for (String part : autoParts) {
    processPart(part);
}
```

The inefficiency in the preceding code stems from the sequential processing of each automotive part within the loop. This can be a time-consuming operation and lead to poor performance. Let's look at an improved version:

```
List<String> autoParts = getAutoParts();
processAutoParts(autoParts);
```

In this improved version, we implement a `processAutoParts()` method that takes the entire parts list as a parameter. This permits batch processing of parts, which can significantly improve overall performance.

With our exploration of loop types completed, let us now review loop optimization basics.

## Loop optimization basics

As we strive to ensure our Java applications are high-performing, we rightly focus on loop optimizations. Loops are equally ubiquitous in Java programming, making their inefficiencies especially problematic. Loop efficiency can vary significantly based on implementation and use. This section examines fundamental aspects of loop optimization, with the aim of equipping you with the knowledge and tools to boost the performance of your Java applications.

### Loop overhead

Loop overhead refers to the additional computational costs associated with the initialization, implementation, and termination of a loop. While loops are essential for achieving various programming tasks, they are not free from computational overhead. Understanding loop overhead is crucial because excessive overhead can degrade the overall performance of your Java applications.

The three main components of loop overhead are loop initialization, condition evaluation, and iteration. As previously illustrated, each type of loop can result in inefficiencies with one or more of these components. Fortunately, there are strategies we can adopt to help minimize loop overhead. First, we can optimize loop initialization. This can be done by initializing our control variables outside of the loop. The performance gained is due to avoiding redundant assignments within the loop. We can also minimize the number of variables declared within the loop.

A second strategy is to use simple and efficient condition expressions. We want to avoid complex conditions within loops. If our loop's condition is dependent on a dataset's length, that length should be cached outside of the loop to prevent repeated access.

Thirdly, we can streamline our iteration. To adopt this strategy, we should ensure the iteration is designed to update loop control variables correctly and efficiently. We also should use the appropriate expression increments/decrements in an efficient and correct manner. Lastly, we should consider alternative loop types for each set of requirements.

### Bottlenecks

Unoptimized loops can result in bottlenecks in the performance of our Java applications. Understanding the cause of these bottlenecks is a necessary first step to avoiding them. As previously covered in this chapter, the following are common performance bottlenecks associated with loops in Java programming:

- Inefficient condition expressions
- Inefficient iteration steps

- Unnecessary computations

- Suboptimal data access patterns

- Excessive memory allocation

We should strive to implement optimization strategies specific to each loop type, as detailed in the preceding sections. General loop optimization strategies are the following:

- Profile our loops to identify performance bottlenecks

- Avoid premature optimization

- Choose appropriate data structures

- Choose appropriate algorithms

Additionally, as you will learn later in this chapter, profiling tools and methodologies can be used to help us determine where bottlenecks are within our loops.

### Benchmarking

Optimizing loops beyond the basic strategies covered in this chapter requires a firm understanding of how they perform, their characteristics, and the ability to measure optimizations for performance. We accomplish this measurement through benchmarking. Our goal is to assess the effectiveness of our loop optimizations.

We should establish a benchmarking environment so that we can evaluate loop performance. This requires us to set up a consistent environment that facilitates accurate measurements and comparative analysis. Each optimization can be tested against the benchmark to determine if it increased or degraded application performance.

To establish a benchmarking environment, follow these steps:

1. Choose a benchmarking framework or library that provides tools for measuring the performance of our Java code. We will take a specific look at the **Java Microbenchmarking Harness (JMH)** tool in *Chapter 13*.

2. Ensure your development environment is properly configured and up to date.

3. Write benchmarking classes that include the loops we want to benchmark.

4. Specify settings and parameters for our benchmarks. This can include measurement time, number of iterations, and warm-up iterations.

5. Ensure that the variables involved are consistent across all tests.

6. Conduct warm-up runs to account for any JVM-specific warm-up effects.

7. Measure execution times.

8. Analyze results.

9.  Repeat and validate multiple times to ensure the validity of your findings.

10. Interpret your results and optimize.

11. Document your results.

By following these steps for establishing a benchmarking environment, you can make informed decisions about your loop optimizations, identify performance bottlenecks, and fine-tune your Java applications to achieve higher performance.

## Loop unrolling

Loop unrolling is an advanced optimization strategy. It is an optimization technique we can employ in Java to improve our loops. The approach involves replicating, or unrolling, the loop's code block multiple times, to reduce the number of required iterations. More specifically, instead of executing our loop's code block once per iteration, we expand the loop to execute the code block multiple times within a single iteration. While this may seem complex, it can reduce the loop overhead caused by condition checks and updating variables.

The benefits of loop unrolling include the following:

- Reduced loop overhead

- Improved CPU instruction cache usage

- Enhanced compiler optimization

- Opportunities for parallelism

There are two types of loop unrolling: manual and automatic. Let's look at an example of each. This first example is of manual loop unrolling. Here, we must explicitly rewrite our loop code to unroll it. Here is an example:

```
// Original loop
for (int i = 0; i < 5; i++) {
  // loop's code block
}
// Manual loop unrolling
for (int i = 0; i < 4; i += 2) {
  // loop's code block (iteration 1)
  // loop's code block (iteration 2)
  // ...
}
```

The manual loop unrolling provides us with finite control over the unrolling process, but, because it is done manually, it is exceedingly difficult to maintain the code.

With automatic loop unrolling, we can leverage compiler optimization tools to automatically unroll our loop code. The compiler will reference compiler flags that we specify. Here is an example:

```
// Original loop
for (int i = 0; i < 5; i++) {
    // loop's code block
    // ...
}
// Compiler optimization flags
```

> **Compiler flags**
>
> In Java, compiler flags are used to specify non-default settings or options at compile time. We pass flags (for example, -O, to enable optimization during compilation) to the `javac` compiler command.

Using automatic unrolling can simplify the optimization process and save us time. The disadvantage to this approach is that the unrolling results might not be the most efficient. As the complexity of our loops increases, the efficiency of the automatic unrolling results decreases.

It is considered a best practice to use loop unrolling when we have a known and fixed number of iterations. Also, if loop overhead is not a significant concern, then loop unrolling may not be necessary. A side effect of manual unrolling may be increased memory usage. Whenever unrolling loops, we should profile and measure optimizations on performance.

There are limitations of loop unrolling. First, not every loop is ideal for unrolling, such as when the dataset size is variable or dynamic. Another limitation or disadvantage of unrolling is that it can result in bloated code, increasing the binary size, which, in turn, can result in instruction cache mistakes. Lastly, automatically unrolling may not always produce the desired loop optimization.

You should now have a comprehensive understanding of loop types in Java, including their strengths and weaknesses and their ideal use cases. Your loop selection decisions can now be more informed and help ensure you are creating code that is performant and maintainable.

## Testing loops for performance

Now that we have a firm grasp of the different types of loops and their advantages and disadvantages, we should feel empowered to make the best loop selections. Furthermore, we should be able to implement optimization strategies. But how do we know if our loop optimization strategies result in better or worse performance? That is where testing comes into play and is the focus of this section.

In this section, we will cover the following concepts:

- Profiling tools and methodology

- Benchmarking and testing strategies

- Case studies and examples

Our goal is to obtain the knowledge and practical skills needed to evaluate, optimize, and harness the full potential of loops within our Java applications. Moreover, we want to have a testing strategy that will inform us of the efficacy of our optimizations.

## Profiling tools and methodology

It is not enough to write optimized loops in our Java applications. We should have a clear understanding of profiling tools and methodologies. In our context, profiling is the practice of analyzing the execution of our loops to gain insights into their performance characteristics. We can glean crucial information on how much CPU time our loops demand.

### Why profiling is important

Profiling is considered a diagnostic tool used by developers who are concerned with the performance of their Java applications. Profiling is applicable to all sections of our code, and, in this section, our focus is on loops. We use profiling tools to get deep insights into the behavior of our code at runtime. Here are some insights we can gain from profiling:

- How much CPU time do our loops consume?

- Are memory leaks present?

- Does our code result in excessive memory allocation?

- How do our loops interact with the CPU cache?

- How do our loops interact with memory?

We can write our loops in different ways and use profiling to help inform us of which method is the best from an overall runtime performance perspective.

### Profiling types

There are three basic profiling types: CPU, memory, and thread. In **CPU profiling**, the focus is on how our code uses the CPU. This profiling type is especially useful for identifying CPU bottlenecks created by our loops.

**Memory profiling** gives us insights into memory-related issues. These can include excessive memory consumption, memory leaks, and excessive or inefficient object creation. This is one of the most useful profiling types when our focus is on loop optimization.

The **thread profiling** type is useful in identifying synchronization issues that can affect the performance of our loops.

## Profiling tools

Profiling tools are software applications that help automate profiling options for us. There are a plethora of tools available to Java developers, many of which are open source and free to use. These tools can be organized into three categories: those that come bundled with the JDK, those that come bundled with an **integrated development environment** (IDE), and commercial or third-party tools.

Let's look at two examples of profiling tools that come bundled with the JDK. First, **Java Flight Recorder** (**JFR**) is a built-in tool that can record and analyze application behavior. This tool is lauded for its low processing overhead. **VisualVM** is another profiling tool that comes bundled with the JDK. It provides great insights into CPU usage and memory usage, as well as thread activity.

## Profiling methodologies

A profiling methodology is essentially your approach to profiling. This is more than simply selecting a profiler to use. In fact, you often will want to use more than one tool as part of your methodology. When asked what your profiling methodology is, consider what your overall approach is.

You might employ a **sampling profiling** methodology where you periodically pull data on your application's execution. This sampling approach provides a quick view of code execution behavior and can provide insights into what might need deeper profiling.

Another common profiling methodology is **instrumentation profiling**. This approach requires us to add profiling code to our application. While this approach can provide the greatest level of fidelity regarding code execution behavior, it also commands the most CPU overhead.

A third popular profiling methodology is **continuous profiling**. This approach collects a lot of data over long periods of an application running. Developers who employ this methodology can gain insights into long-term trends and more easily detect performance anomalies.

> **Identifying hotspots in our loops**
>
> In the context of Java loops, a hotspot is a loop that consumes excessive CPU time or memory at runtime. It is imperative that we pinpoint these hotspots in our loops so that we can optimize them.

As a final note on profiling, we should realize that identifying a **hotspot** is just one step, and using a profiling tool is another. We need to address the issues we uncover and then reapply profiling tools to ensure our changes have the desired impact. Developers should have a continuous improvement mindset over their loops and other code. This requires the continual monitoring and profiling of our code.

## Benchmarking and testing strategies

We have accepted that optimizing loops is a critical part of ensuring our Java applications perform at a high level. This high performance requires benchmarking and testing strategies. This section looks at both types of strategies and shares best practices.

### Benchmarking strategies

JMH is the most common toolkit used for conducting performance testing of Java code. It does a wonderful job of handling finite components of Java performance testing. As the name suggests, it is designed specifically to test Java code and is widely used in micro-level performance testing. Some features of JMH are the following:

- It measures the performance characteristics of microbenchmarks (code segments). It is especially handy for loop performance testing.
- JMH supports measurements of average time, sample time, single-shot time, and throughput. This gives developers great flexibility.
- With JMH, developers have finite control of the testing environment.
- Can be integrated with build tools (for example, Maven).
- Provides detailed microbenchmarking results.

Regardless of which benchmarking tool we use, it is important to first establish a baseline. We can accomplish this by running our tools on unoptimized code. Once we have the results of that test, we have our baseline. Testing optimized code can be evaluated against the baseline.

### Testing strategies

There are three primary testing strategies. The first is unit testing. This strategy can help us ensure that any changes we make to our loops do not impact the expected code behavior. Unit testing can also help us test our loops with edge cases such as with extreme values.

Profiling and hotspot analysis is another testing strategy that has already been covered in this chapter. As a reminder, we use profiler tools such as JFR or VisualVM to help us analyze CPU and memory usage.

Lastly, we can use regression testing to test our loops after we make changes to the application. This can help ensure that our changes do not negatively impact any other functionality.

Regardless of our testing strategy, we should approach loop optimization testing in an iterative manner. This means we should make small, incremental changes to our loops, testing after each change.

# Case studies and examples

This section covers practical case studies and examples of optimizing loops in Java. It can be powerful to review real-world scenarios as we stand to gain great insights into common challenges and best practices for solutions.

## Case studies

Consider a case study in which we need to process a large dataset and aggregate data based on a business requirement. Our challenge, in this scenario, is that our aggregation loop is very slow. Since we have a large dataset, we want to solve for the slow loop. What can we do? We can ensure we are using an `ArrayList` instead of a `LinkedList`. We can also use Java's **Stream API** to implement parallel processing. This should make better use of multi-core CPUs. We can also minimize object creation within the loop. Given this scenario and the suggested solutions, we are likely to significantly reduce the aggregation loop operation time as well as the complexity of the loop.

Let's look at another case study. This one features a banking application that calculates investment risk metrics over large datasets. The challenge, in this scenario, is that our main calculation loop is inefficient. We might take a three-pronged optimization strategy. First, we will examine our algorithm and see if it can be improved. Next, we will write our code so that it supports **Just-In-Time (JIT)** compilation. Lastly, we will look at our data structures and make any necessary changes to minimize memory access issues. This three-pronged approach can result in a significantly more performant application.

## Examples

Let's look at two real-world applications of loop testing and optimization.

Our first example is data processing for healthcare analytics. The challenge is that processing large groups of patient data for analytical reports is slow. The solution is to implement multithreading and batch processing within our loops so that we can process data in parallel chunks. The result of this approach should be a significant reduction in data processing time.

Another example is inventory management as part of an e-commerce Java application. The challenge is to optimize the loop in our code that processes inventory updates. For this scenario, we will assume that the loop is slower than it has been in the past, most likely due to an ever-increasing inventory dataset. The solution would likely be to use efficient data structures and change our loop structure as appropriate. The result of this approach should be faster inventory processing and greater response times.

This section covered profiling tools, profiling methodologies, benchmarking, testing strategies, case studies, and examples. We can now confidently evaluate, optimize, and harness the full potential of loops within our Java applications, leading to high runtime performance.

# Nested loops

Now, we'll cover the fundamental concepts, practical optimization strategies, and technical intricacies involved in effectively optimizing nested loops in Java applications. We will tackle the concept of nested loops, as they relate to high-performance Java applications, in the following sections:

- Introduction to nested loops
- Loop fusion in nested loops
- Parallelizing nested loops
- Nested loop vectorization

It is important to understand when to use nested loops and, when we do, how to implement them in the most optimal method.

## Introduction to nested loops

A nested loop is when one loop is located inside another. This creates a complex iteration scenario. Here is the syntax for a simple nested loop:

```
for (int i = 0; i < 10; i++) {          // Outer loop
    for (int j = 0; j < 10; j++) {      // Inner loop
        System.out.println("i = " + i + ", j = " + j);
    }
}
```

As you can see in the preceding syntax, the outer loop runs 10 times, and for each iteration, the inner loop also runs three times. This results in a total of 10 x 10 = 100 iterations.

We can implement a finite number of levels in our nested loops. With each inner loop, our code becomes more complex, more difficult to read, and more frustrating to maintain. So, when would we use nested loops? One common implementation is when processing multi-dimensional arrays such as matrices. They are also used when performing operations on tables.

> **Warning**
> Be careful when implementing nested loops. They can very quickly become inefficient and make your application sluggish and non-responsive.

## Loop fusion in nested loops

If we must implement nested loops in our Java applications, we should consider applying loop fusion to them for efficiency. Loop fusion is essentially an algorithmic approach where we combine adjacent loops, ones that perform operations on the same dataset, into a single loop.

We can also attempt to reduce redundant calculations and improve cache utilization. This is possible when we merge loops because it stands to minimize redundancy, thereby improving the use of cache. The technique involves reusing data that is loaded into the cache during the same iteration. The outcome is generally fewer cache misses.

Loop fusion can also prove beneficial when using complex algorithms such as matrix multiplication. There is a potential side effect, namely increased loop complexity. This greater complexity results in decreased code readability and makes the code more difficult to maintain.

## Parallelizing nested loops

To parallelize nested loops in Java, we need to restructure our loops so they can execute concurrently across multiple CPU cores instead of on a single core. As you would expect, this is a component of parallel computing, which can result in significantly increased runtime performance. Strategies for parallelizing nested loops include chunking our dataset and processing each chunk in a separate thread.

The concept of multithreading is often discussed along with parallel streams in the context of nested loops. Comparing the two, we can learn that multithreading offers additional control but requires us to explicitly manage threads and tasks. Parallel streams, which are part of Java's **Stream API**, provide us with an easier method to parallelize operations, but with less control over the threading model.

Lastly, when parallelizing, we must handle synchronization and thread safety. Synchronization becomes key as does ensuring thread safety. This is especially true when we deal with shared data structures.

## Nested loop vectorization

You may recall from *Chapter 2* of this book that vectorization refers to the process of executing a single instruction simultaneously on multiple data points. We can leverage vectorization for nested loops to significantly increase computational speed, especially when dealing with complex data manipulations such as a matrix.

**Single Instruction, Multiple Data (SIMD)** can be an important concept in nested loop optimization. SIMD is a key concept in vectorization where, as the name suggests, a single operation is performed on multiple data elements. This is an especially effective technique with repetitive nested loops. The efficacy of vectorization depends on compiler optimizations and hardware support. Not all CPUs have SIMD capabilities.

This section provided a structured overview of nested loop optimization with the goal of increasing Java application performance. Fundamental concepts were covered, as were practical optimization strategies, and some technical intricacies involved in effectively optimizing nested loops in Java applications.

## Summary

This chapter focused on loops, fundamental programming constructs, and how to get the most out of them from a runtime performance perspective. Concepts covered included loop overhead, loop unrolling, benchmarks, loop fusion, loop parallelization, and loop vectorization. We used code examples to provide insights and demonstrate best practices.

We explored techniques, strategies, and best practices to help you get the best performance out of your loops and to prevent the unintentional introduction of substantial bottlenecks that undermine runtime performance.

The next chapter takes a specific look at Java object pooling, which is a design pattern used to manage reusable objects to conserve resources and improve application performance.

# 4

# Java Object Pooling

Our mission to make our Java applications highly performant includes a look at Java **object pooling**. This chapter dives into the concept of object pooling in Java and how to achieve high performance with them in your Java applications.

The chapter begins with an explanation of object pooling and how to implement an **object pool** in Java. Sample code is provided to help you understand object pooling operations specific to the Java programming language. You will also have the opportunity to learn about the advantages and disadvantages of object pooling in Java. Finally, the chapter shows how you can implement performance testing with Java object pools.

This chapter covers the following main topics:

- Jumping into the object pool
- Advantages and disadvantages
- Performance testing

By the end of this chapter, you should have a strong theoretical understanding of Java object pooling as well as hands-on implementation experience. This experience can help ensure you get high performance out of your Java applications.

## Technical requirements

To follow the examples and instructions in this chapter, you will need the ability to load, edit, and run Java code. If you have not set up your development environment, please refer back to *Chapter 1*.

The finished code for this chapter can be found here: `https://github.com/PacktPublishing/High-Performance-with-Java/tree/main/Chapter04`.

# Jumping into the object pool

Before we jump into the object pool, let's look at what an object pool is.

> **Object pool**
> An object pool is a collection (pool) of objects that can be reused.

Using object pools is an optimization approach that can positively impact the performance of an application. Instead of recreating objects every time we need them, we pool a collection of objects and simply recycle them. To help understand object pooling, consider a real-world example of a physical library. The library can lend out books (our objects) and return them to the collection (our pool) when the person is done with the book. This allows the library to reissue the book to the next person that needs it. Consider the alternative. If the library destroyed (garbage collection) the book after each use, it would have to create a new one each time it is needed. This would not be efficient.

## Database example

A common object pooling implementation in Java programming is with database connections. The typical approach to database connections is to open a connection to the database and then perform the desired operations to update or query the database. The **open-query-close** process is used. The problem with this approach is that opening and closing databases frequently can impact the overall performance of the Java application. This processing overhead is something we should try to avoid.

The object pooling approach, with our database example, involves maintaining a pool of pre-created database requests that are idle. When the app signals a request for a database connection, one is used from the pool. The next section demonstrates how to create and use these pools in a Java application.

## Implementing an object pool in Java

Implementing an object pool, based on the database connection example from the previous section, involves a database connection class, an object pool class, and a class that contains a main() method. Let's look at each of these individually.

> **Note**
> This example simulates object pooling and does not connect to an actual database.

First, we have our DBConnect class. This is the class that we will pool:

```java
public class DBConnect {
  static int dbConnectionCount = 0;
  private int dbConnectID;
```

```
  public DBConnect() {
    this.dbConnectID = ++dbConnectionCount;
    System.out.println("Database connection created: DBConnect" +
    dbConnectID + ".");
  }

  public void dbMethod() {
    // placeholder
  }

  public void dbConnectionClose() {
    System.out.println("Database connection closed: DBConnect" +
    dbConnectID + ".");
  }
}
```

As you can see by the preceding code, there are placeholders for functionality.

Next, we create a `DBConnectObjectPool` class to maintain a collection (pool) of `DBConnect` objects:

```
import java.util.LinkedList;
import java.util.Queue;
public class DBConnectObjectPool {
  private final Queue<DBConnect> pool;
  private final int maxSize;

  public DBConnectObjectPool(int size) {
    this.maxSize = size;
    this.pool = new LinkedList<>();
  }
  public synchronized DBConnect getConnection() {
    if (pool.isEmpty()) {
      if (DBConnect.dbConnectionCount < maxSize) {
        return new DBConnect();
      }
      throw new RuntimeException("Error: Maximum object pool size
      reached. There are no DB connections available.");
    }
    return pool.poll();
  }
  public synchronized void releaseConnection(DBConnect dbConnection) {
    if (pool.size() < maxSize) {
      pool.offer(dbConnection);
      System.out.println("Splash: Connection object returned to the
      pool.");
```

```
      } else {
        dbConnection.dbConnectionClose();
      }
    }
  }
}
```

As shown in the preceding code, we assume a maximum number of connections. This is considered a best practice.

Lastly, we have a partial application to demonstrate how to use our object pool:

```
public class ObjectPoolDemoApp {
  public static void main(String[] args) {
    DBConnectObjectPool objectPool = new DBConnectObjectPool(8);
    for (int i = 0; i < 10; i++) {
      DBConnect conn = objectPool.getConnection();
      conn.dbMethod();
      objectPool.releaseConnection(conn);
    }
  }
}
```

When our application requests a database connection, one is provided from the pool. In the case when a connection is not available from the pool, a new one is created. We do check to ensure that we do not exceed the maximum number of allowed connections. Lastly, after a DBConnect object is used, it is returned to the object pool.

# Advantages and disadvantages of object pooling

Now that you understand what object pooling is and how to implement it in Java, we should consider if this is the right strategy for our application. With most application code optimization approaches, there are both advantages and disadvantages. This section looks at both.

## Advantages

There are several potential advantages of using object pooling. These advantages can be grouped into performance, resource management, and scalability categories.

### Performance advantages

Implementing object pooling stands to allow us to avoid the overhead from object creation. This approach is especially useful in high transaction applications and when system response time is important. Through object pooling, we can help ensure our Java applications are able to be more performant by avoiding excessive object creation.

We can also experience consistent performance between app usage. For example, using object pooling should result in consistent app performance with both a minimal load and a heavy load. This predictable behavior is possible due to the stability of our application. That stability is enabled by avoiding frequent object creation and a heavy reliance on garbage collection.

### Resource management advantages

In the context of object pooling advantages, a resource refers to real time, processing load, and memory. Reducing the number of object creation and destruction operations is a benefit of the object pooling approach. The example used earlier in this chapter was with database connections. That example was used because database connection operations are notoriously resource hogs. The object pooling approach reduces the time it takes to perform these operations and is less resource intensive.

Another resource management advantage is that it increases our memory management schema. When object creation is not controlled, the amount of memory consumed is variable and could result in system errors.

### Scalability advantages

The third category of advantages is the ability for our applications to be more scalable. This is especially true when we have applications with a large number of simultaneous users. It is also beneficial when dealing with database connections where the database is a shared resource. The object pool essentially serves as a buffer for those requests.

Another reason our applications that use object pooling are more scalable is the increased amount of control we have with our resources. In the database connection example presented earlier in this chapter, we set the maximum number of objects that could be in the pool.

## Disadvantages

Unfortunately, there are more potential disadvantages to using object pooling than there are advantages. These disadvantages can be grouped into code complexity and resource management categories.

### Code complexity disadvantages

Like with any non-standard programming approach, object pooling adds complexity to our code. We create our object pooling-related classes, which must contain algorithms to manage the object pool and interfaces with the main program. Although it is not apt to result in bloated code, it can make it difficult to maintain.

Object pooling, when implemented in a Java application, adds another component to test each time the system, connected systems, or data changes. This can be time and resource intensive.

### Resource management disadvantages

There is always a risk, especially during peak load times, of having enough available resources. When we set a maximum size of our object pools, they might not be sufficient to handle those peak load times. This can also be referred to as **resource starvation** because all objects in our pool have been allocated, preventing new requests from being queued. These delays can result in overall performance lag and user dissatisfaction.

Working with memory allocation and deallocation can be problematic. If we do not, for example, manage how objects are returned to the pool, there could be data loss. This could compound a situation where there are no objects available in the pool. Implementing error checking and exception handling becomes critical.

Finally, we need to maintain the balance of an object pool being too large or too small. If it is too small, it could result in extensive queue times for pooled objects. If the pool is too large, the application might overconsume memory, taking away from other areas of the application that could make use of it.

After looking at both the advantages and disadvantages, you should be able to determine whether object pooling is ideally suited for your application.

# Performance testing

When we implement object pooling in our Java applications, we want to do three things:

- Ensure our program works
- Prove that our implementation resulted in greater performance
- Quantify the optimization

In previous sections, we looked at how to implement object pools in Java. In this section, we will look at how to design a performance test, how to implement the object pooling performance test, and how to analyze the testing results.

## Designing a performance test

After we decide to implement a performance test, our first action is to design the test. The questions we need to answer here include the following:

- What is our goal?
- What will we measure?
- How will we measure?
- What conditions will exist for our tests?

With these questions in mind, we can start designing our performance test. We should have a clear goal or set of goals for our performance test. We might, for example, want to focus on system memory, CPU load, and so on.

Once we have a specific goal, we must decide what to measure. In testing, what we measure are considered **key performance indicators** (**KPIs**). The performance testing of object pools might be memory usage, CPU use, data throughput, and response time. These are just some examples.

Next, we will need to set up our test environment and create test scenarios. The test environment closely replicates the production system. You might duplicate your system in a development environment, so the live system is not impacted. Likewise, the test scenarios should closely resemble the real-world use of your system. To the extent possible, we should create as many different scenarios as needed to represent what our live system handles.

At this point, you are ready to document your test plan and implement it. The next section covers how to implement a performance test.

## Implementing a performance test

Implementing your test plan should not be terribly difficult. Here, you are simply putting your plan into action. The test environment is established, and you run your test scenarios. As the test is running, you should be collecting the data for later analysis. Of critical importance is the ability to reproduce your test conditions to support future comparative tests.

Let's look at how a performance test might be written in Java using the database connection example from this chapter. We will set the goal of decreasing the time our application takes to obtain a database connection from our object pool and perform a simple operation on that database. Our test plan will compare the results of our test with the same test on a version of our application that does not implement an object pool.

Our code starts with the class declaration and class variables:

```
public class DBConnectionPerformanceTest {
   private static final int NUMBER_OF_TESTS = 3000;
   private static DBConnectObjectPool dbPool = new
   DBConnectObjectPool(24);
```

Next, we will write the first part of the main() method. This first snippet of code will be how we perform the test with our object pool:

```
public static void main(String[] args) {
   long startTime_withPooling = System.nanoTime();
   for (int i = 0; i < NUMBER_OF_TESTS; i++) {
     DBConnect conn = dbPool.getConnection();
     conn.dbMethod();
```

```
    dbPool.releaseConnection(conn);
    }
    long endTime_withPooling = System.nanoTime();
```

Now, we will write the code to test without using the object pool:

```
long startTime_withoutPooling = System.nanoTime();
    for (int i = 0; i < NUMBER_OF_TESTS; i++) {
        DBConnect conn = new DBConnect();
        conn.dbMethod();
        conn.dbConnectionClose();
    }
    long endTime_withoutPooling = System.nanoTime();
```

With both sets of performance testing written, we need to add the ability to calculate and output the results. We generate the results by simply subtracting the startTime value from the endTime value and converting it to milliseconds. We then output the results to the console:

```
long totalTime_withPooling = (endTime_withPooling - startTime_
withPooling) / 1_000_000;
long totalTime_withoutPooling = (endTime_withoutPooling - startTime_
withoutPooling) / 1_000_000;
System.out.println("Total time with object pooling: " + totalTime_
withPooling + " ms");
System.out.println("Total time without object pooling: " + totalTime_
withoutPooling + " ms");
```

This simple example of an object pool performance test is intended to give you a general idea of how to code these tests. Every application is different and how you write your performance tests will vary.

## Analyzing the results

Once our tests have concluded, we can analyze the results. How you analyze the results will depend on your goals and KPIs. The analysis task should not be rushed. Remember, you collected this data so it could help inform your decision on your object pool. The complexity will vary based on the performance test plan.

Given the database connection example, we can simply add it to the bottom of our DBConnectionPerformanceTest class to compare the two sets of results. Here is the first section of that code:

```
if (totalTime_withPooling < totalTime_withoutPooling) {
    System.out.println("Results with object pooling: " + tctalTime_
    withPooling);
    System.out.println("Results without object pooling: " + totalTime_
    withoutPooling);
```

```
    System.out.println("Analysis: Object pooling is faster by " +
    (totalTime_withoutPooling - totalTime_withPooling) + " ms");
}
```

As you can see, we simply check to see whether `totalTime_withPooling` is less than `totalTime_withoutPooling`. If this is the case, the relevant results are displayed on the console.

Next, we will check to see whether `totalTime_withPooling` is greater than `totalTime_withoutPooling`. The relevant results are displayed on the console:

```
} else if (totalTime_withPooling > totalTime_withoutPooling) {
    System.out.println("Results with object pooling: " + totalTime_
    withPooling);
    System.out.println("Results without object pooling: " + totalTime_
    withoutPooling);
    System.out.println("Analysis: Object pooling is slower by " +
    (totalTime_withPooling - totalTime_withoutPooling) + " ms");
}
```

Our final code snippet executes when the first two conditions are not met. This means that both tests took the same amount of time:

```
} else {
    System.out.println("Results with object pooling: " + totalTime_
    withPooling);
    System.out.println("Results without object pooling: " + totalTime_
    withoutPooling);
    System.out.println("Analysis: No significant time difference between
    object pooling and non-pooling.");
}
```

As with all testing, you should document your plan, the test results, your analysis, your conclusions, and your actions following the testing. This robust documentation approach will help you retain the history of your testing in detail.

## Summary

This chapter took an in-depth look at Java object pooling. It was suggested that object pooling is an important technique for ensuring our Java applications perform at a high level. Armed with theoretical knowledge, the chapter explored the advantages and disadvantages of object pooling. We focused on areas such as memory, CPU use, and code complexity. Finally, we demonstrated how to create a performance testing plan, how to implement it, and how to analyze the results.

In the next chapter, we will focus on algorithm efficiencies. Our goal will be to ensure our algorithms have low time complexities. The chapter will demonstrate inefficient algorithms and how to transform them to support high performance.

# 5

# Algorithm Efficiencies

Developers focused on ensuring their Java applications perform at a high level must consider the efficiency of individual algorithms. We do not judge an algorithm's efficiency by its lines of code; rather, we make this judgment after analyzing test results.

This chapter is intended to help you learn how to choose the right algorithm for any given requirement. It also covers the concept of time complexity to include strategies for reducing time complexity. We will also focus on lean and efficient code. This chapter also emphasizes the importance of algorithm testing.

This chapter covers the following main topics:

- Algorithm selection
- Low time complexity
- Testing algorithms for efficiency

By the end of this chapter, you should have a strong theoretical understanding of algorithm efficiencies as well as hands-on experience creating and modifying algorithms. This experience can help ensure you get high performance out of your Java applications.

## Technical requirements

To follow the examples and instructions in this chapter, you will need the ability to load, edit, and run Java code. If you have not set up your development environment, please refer to *Chapter 1*.

The finished code for this chapter can be found here: https://github.com/PacktPublishing/High-Performance-with-Java/tree/main/Chapter05.

# Algorithm selection

One of the best things about writing software is that there is no single solution to a problem. We are free to use a style that's unique to us and to incorporate data structures, libraries, and algorithms as long as we obtain the correct results. This is a bit of an overstatement. We can write an algorithm in essentially infinite different ways and get the same results.

This programming flexibility can also be a detriment, as evidenced by low-performant algorithms. So, just because we can write algorithms with reckless abandon, it does not mean we should. We should be strategic in our algorithm selection and creation as it will have a significant impact on the overall efficiency of our Java applications.

In this section, we will look at a specific process of selecting algorithms, a case study, and evolving trends.

## Selection process

While there is no industry-wide official algorithm selection process, here is a six-step approach:

1.  Fully understand the requirements. It may seem obvious, but so many developers get this wrong. It is critical to fully understand the requirements. This is the problem we want an algorithm to solve. Understanding the problem involves learning about constraints, datasets, inputs, outputs, and more. Once we have fully understood the requirements, we can move on to the next step.

2.  In this second step, we should become familiar with the data that our algorithm will be associated with. This might include static data, data streams, or even generating data. This step is where you fully immerse yourself in anything related to the application's data. In addition to data familiarity, we should start thinking about what type of data structures our algorithms should use.

3.  Consider the computational complexity of each potential algorithm. Points for consideration here include the time and memory requirements. You can implement benchmarking to help inform your decision.

In this step, you should identify resource limitations related to processing and memory. Knowing these upper boundaries will aid your algorithm decision.

This is where you test your algorithms, make refinements, and retest. This is an iterative process where you should make small, incremental changes to your algorithms to help identify what works best.

Document your final decision and all factors that led to it. The more detailed your document, the better. This will be helpful when revisiting your selection later.

Ideally, you should adopt a continual process improvement mindset and not hesitate to question previous decisions. Even if you select and refine an algorithm that results in the best possible performance, it should be periodically reviewed. Environments, data, and other factors can change over time and impact your algorithm's efficiency.

Now, let's see how these steps can be implemented by reviewing a case study in the next section.

## Case study

It can be helpful to review case studies to help solidify our understanding of how algorithm selection can significantly improve Java application performance. This section details the algorithm selection for a business' e-commerce platform. The business noted that its system processed search requests inefficiently, resulting in user complaints that the system was sluggish. So, the business hopes to improve the search function and, therefore, the overall performance of their application.

The first step in this case is to identify what is causing the problem. For our example, the e-commerce platform implemented a linear search algorithm that matches user requests with database entries. The search algorithm has not changed since the initial system was developed and the database continues to increase in size. All search results are slow.

Our second step is to understand the data. So, the developer reviews the database schema and sampled database records to become familiar with the system's data requirements and use. In the next two steps, the developer reviews the computational complexity of the search feature and identifies any resource limitations.

The developer is now poised to evaluate alternative algorithms. In this example, possible alternatives could include binary searches and other proven search patterns. The review of these alternatives leads to the adoption of an inverted index for a data structure that can be optimized for full-text searches.

> Inverted index
>
> A data structure commonly used for full-text searches. The indexing identifies each unique term and lists where that term appears. Implementation typically includes using a dictionary or hashmap.

Now the developer can tailor their selection based on their system's needs. Next, they will perform multiple tests and document the results. In this example case, the system's search function is now highly performant. What is left is to document the decision and decision factors.

The example case showed developers that they need to balance application performance and resource utilization. Next, let's look at some evolving trends in algorithm selection.

## Evolving trends

Java software development is a dynamic field, as are the algorithms we create and use. There are several trends regarding algorithm efficiencies, and five of them are described in this section.

- **AI and ML**: Many people in the tech space have coined 2023 as the year of **Artificial Intelligence** (**AI**). AI and **Machine Learning** (**ML**) are not new, but their widespread use and adoption have never been more prevalent, thanks to OpenAI's launch in 2015 and the introduction of **Generative Pre-trained Transformers** (**GPTs**), such as ChatGPT. The industry is experiencing a shift towards AI-related algorithms.

You can learn more about leveraging AI for high-performance Java applications in *Chapter 18*.

- **Concurrency**: With the proliferation of multi-core processors, the need to focus algorithms on parallel processing has increased. Fortunately, Java has robust support for concurrency.

  You can learn more about concurrency strategies in *Chapter 9*.

- **Cloud optimizations**: As more applications are being hosted in the cloud, the need for updating our Java applications for cloud optimization increases. Cloud computing requires us to consider concepts such as scalability and distributed processing. These considerations are important for new Java applications that are being developed for the cloud as well as existing applications that we want to migrate to the cloud.

- **Security**: One thing in software development that seems to remain the same is the ever-increasing number of cyber threats to our systems and data. Security is a mindset, not a step in software development. The trend here is the increased emphasis on security throughout the product life cycle.

- **Community**: There is impressive growth in the area of crowdsourced algorithm design and other open source contributions. The systems and algorithms that result are increasingly robust and diverse.

Now that we have reviewed algorithm selection, shared a case study, and discussed evolving algorithm trends, we can look at low time complexity as it relates to algorithm efficiencies.

# Low time complexity

Time complexity refers to the measure of the time efficiency of any given algorithm. We want to determine the execution time of our Java applications and ensure our algorithms do not add time complexity to our applications. The overall goal is to reduce algorithm execution time. We should test our algorithms using a variety of possible inputs and environments.

In this section, we will look at specific strategies to help you reduce the time complexity of your algorithms. We will also address the common pitfalls associated with time complexity.

## Strategies for reducing time complexity

There are several strategies we can adopt to help us reduce the time complexity of our algorithms. Perhaps the easiest strategy is to simply ensure our algorithms are not overly complicated. This strategy requires us to double-check our algorithmic logic and use the optimization techniques covered throughout this book. Reducing unnecessary calculations within a loop, especially with recursion, is a great optimization strategy.

Sometimes our algorithms might seem to be overly complex and we cannot find a way to simplify them. In those cases, it is worth considering breaking the algorithm into multiple algorithms, each executing a subset of the original set of instructions.

## Common pitfalls

When reviewing the common pitfalls related to algorithm time efficiency, we can refer to previous chapters and look at others yet to come. As you can see from the list of the top four common pitfalls listed here, details can be found in several chapters of this book:

- Improper use of data structures (*Chapter 2*)

- Overuse of non-standard libraries (*Chapter 13*)

- Inadequate testing and profiling (*Chapter 14*)

- Lack of focus on readability and maintainability (*Chapter 16*)

In addition, we might simply be overlooking the big picture of our application's overall performance. Sometimes, we can get stuck in the minutiae of our code at the expense of overall algorithm or application efficiency.

As we have seen, low-time complexity is both a concept and a goal. It is an essential consideration when our ultimate aim is to increase the performance of our Java applications.

# Testing algorithms for efficiency

At this point, it is clear that we need our algorithms to be efficient. This section focuses on how we can measure algorithm efficiency based on our requirements. We will start with a short section on the importance of testing, then how to prepare for testing, how to conduct the tests, and what to do after testing.

## Importance of testing

Simply put, if we do not test our algorithms for efficiency, we will not know for sure if they are efficient or inefficient. Optimizations are pointless if we do not measure their impact on a specific algorithm and overall application performance.

We might assume that we know what optimizations will result in the best performance based on our experience. Although this is experience-based, it is still merely anecdotal. Changes in input and the operating environment can challenge our previous knowledge of algorithm efficiency, so we should not assume without testing.

> **Tip: Become a testing zealot**
> Good programmers focused on high performance are dedicated to the idea that algorithm efficiency testing is a core part of the software life cycle.

## Preparing for algorithm efficiency testing

It can take time to set up a testing environment, so it is important to plan for it and allot sufficient time in your development project for it. Here is an overview of a seven-step process of preparing for algorithm efficiency testing:

1. **Identify requirements**: In this first step, we need to thoroughly understand what the algorithm is intended to do and what **Key Performance Indicators** (**KPIs**) you will use in your testing.

2. **Establish the test environment**: We rarely test in a production environment, so we need to replicate data and conditions in the development environment. The goal is to mirror, as closely as possible, the production environment.

3. **Tool selection**: In this step, we select the profiling and benchmarking tools we will use.

4. **Obtaining test data**: Here, we want to create a replica of live data so we can use it for testing.

5. **Benchmarking**: In this step, we establish baselines so that we can compare our current algorithms to future, optimized versions.

6. **Thorough documentation**: It is important to document our testing plans so we can replicate them in the future.

7. **Iterative testing**: This final step is essentially a scheduling step. In order to support the need for iterative testing, multiple iterations should be scheduled.

Now that you know how to prepare for testing, let's look at how to conduct the tests.

## Conducting the tests

You are now ready to implement your testing plan. Let's look at a simple example of what this looks like in Java code.

Our example will be to test a bubble sort algorithm's execution time. Here is our bubble sort algorithm:

```java
public class EfficiencyTestExample {
  public static void main(String[] args) {
    algorithmEfficiencyTest();
  }

  // Here is our bubble sort algorithm
  public static void ourBubbleSort(int[] array) {
    int nbrElements = array.length;
    int temp;
    for (int i = 0; i < nbrElements; i++) {
      for (int j = 1; j < (nbrElements - i); j++) {
        if (array[j - 1] > array[j]) {
          temp = array[j - 1];
```

```
            array[j - 1] = array[j];
            array[j] = temp;
        }
      }
    }
  }
}
```

Next, we need to create a method for testing the execution time of our bubble sort algorithm. This is what that would look like:

```
public static void algorithmEfficiencyTest() {
  int[] ourTestArray = new int[30000];
  for (int i = 0; i < ourTestArray.length; i++) {
    ourTestArray[i] = (int) (Math.random() * 30000);
  }
  long startTime = System.nanoTime();
  ourBubbleSort(ourTestArray);
  long endTime = System.nanoTime();
  long duration = (endTime - startTime);
  System.out.println("Execution Time (nanoseconds): " + duration);
}
```

As you can see, we created a test array with 30,000 elements and then populated it with random numbers. Next, we logged our start time, executed the bubble sort, and then logged the end time. With both the start and end times, we know how long the algorithm took. This is what the output might look like:

```
Execution Time (nanoseconds): 1054427380
```

Next, let's look at what our post-testing actions should be.

## Post-test actions

Our work does not end after we complete our tests. We need to perform the following post-test actions:

- Thoroughly analyze our test results
- Optimize our algorithms based on our test results
- Document changes to our algorithms and record why we made the changes
- As appropriate, share your findings and process with stakeholders
- When necessary, update our test environments
- Adopt a continuous process improvement mindset with constant monitoring
- Conduct an after-action reflection and document what you learned about the process and how it can be improved

When we thoroughly analyze our test results and conduct the other post-test actions, we increase the chance of our testing plans being properly executed and with valid results.

## Summary

This chapter was intended to help you learn how to choose the right algorithm for any given requirement and how to measure your results. It covered the concept of time complexity and included strategies for reducing time complexity. We also emphasized the importance of algorithm testing. You should now have a strong theoretical understanding of algorithm efficiencies as well as hands-on experience of creating and modifying algorithms. This experience can help ensure you get high performance out of your Java applications.

In the next chapter, *Strategic Object Creation and Immutability*, we will introduce strategies for object creation with the mindset of improving the overall performance of our Java applications. That chapter covers minimizing object creation, object immutability, and garbage collection.

# Part 2: Memory Optimization and I/O Operations

Memory management is crucial for high-performance Java applications. This part focuses on strategic object creation and the use of immutability to optimize memory usage. It also covers the effective handling of string objects and the identification and prevention of memory leaks. The chapters in this part provide practical insights into managing memory efficiently and ensuring robust I/O operations.

This part has the following chapters:

- *Chapter 6, Strategic Object Creation and Immutability*
- *Chapter 7, String Objects*
- *Chapter 8, Memory Leaks*

# 6

# Strategic Object Creation and Immutability

This chapter continues our quest to find ways to get the most performance out of our Java applications. Creating objects is a core part of all Java applications, so the goal is not to eradicate that; rather, it is to take a strategic approach to object creation.

How and when we create objects can play a crucial role in application performance. Object creation impacts not only performance but also overall efficiency, garbage collection, and memory use. This chapter intends to provide you with the knowledge and skills you'll need to implement an object creation strategy.

A core part of object creation strategies is the concept of object immutability. This chapter presents information and examples on how to make objects immutable and explains why you should consider it.

This chapter covers the following main topics:

- Minimizing object creation
- Object immutability
- Garbage collection
- Design patterns

By the end of this chapter, you should have an understanding and appreciation for the importance of strategic object creation and the powerful concept of object immutability. This understanding will help you improve the performance of your Java applications.

## Technical requirements

To follow the examples and instructions in this chapter, you will need to be able to load, edit, and run Java code. If you haven't set up your development environment, please refer to *Chapter 1*.

The code for this chapter can be found here: `https://github.com/PacktPublishing/High-Performance-with-Java/tree/main/Chapter06`.

# Minimizing object creation

Minimizing object creation is a critical concern when we are striving for high-performance applications. Every time we create an object, we use memory and processing resources. Although modern systems have an impressive array of memory capacity and processing capabilities, they are not limitless.

To ensure we handle this concern correctly, we should seek to understand the life cycle of a Java object, how object creation impacts memory, and what object pooling is. We should also experiment with different object initialization approaches and ways to reduce system overhead. That is the aim of this section. Let's start our exploration of minimizing object creation by looking at the life cycle of Java objects.

## Java object life cycle

The notion of an object in the Java programming language is nothing new. When considering high-performance Java applications, we need to consider the overall life cycle of our objects. The creation, use, and deletion of objects have a direct impact on the overall performance of our applications. The following figure depicts a typical object's life cycle:

Figure 6.1 – Java object life cycle

As illustrated in *Figure 6.1*, the first step of an object's life cycle is object creation. When a new object is created, it is said to be **instantiated**, meaning that we created an instance of the class. This is also referred to as **instantiation**. As you can see from the following code, we use Java's new keyword to allocate memory on the heap so that we can store our object's data:

```
public class Corgi {
  // Instance variables
  private String name;
  private int age;
  private int weight;

  // Constructor
  public Corgi(String name, int age, int weight) {
    this.name = name;
    this.age = age;
```

```
    this.weight = weight;
  }

  // Getter methods
  // Setter methods
  // Methods
}
```

The next phase of an object's life cycle is its use and reference. This is where we execute the object's methods and reference its properties.

The third phase is garbage collection. Once we stop using an object and it can no longer be referenced, the **Java Virtual Machine's (JVM's)** garbage collector will reclaim the memory the out-of-scope object is using.

The last phase of an object's life cycle is when it is destroyed. The garbage collector takes care of this for us. Once the object is destroyed, it is no longer accessible to the application.

Understanding an object's life cycle is a prerequisite to being able to create and adopt an object creation strategy.

## Memory

We implicitly understand that objects require memory to exist and that the more objects our applications use at one time, the more memory that is required. To support high performance, we should strive to understand how Java manages memory specific to object creation. To aid in our understanding, let's look at three.

### Stack versus heap memory

**Stacks** and **heaps** are both used for memory allocation, are used for different purposes, and behave differently. Let's start by defining a stack.

---

Stack

A stack is the region of memory where a static collection of elements using a **last in, first out (LIFO)** model is stored.

---

Next, let's look at what a heap is so that we can compare it to a stack and determine how both of them impact performance.

---

Heap

A heap is an area of memory that's used for dynamic memory allocation. Objects are allocated on the heap when we use the new keyword in our applications.

---

Java uses both stacks and heaps and, as you've learned, they are used differently. Let's look more deeply at stacks. We typically use stacks to store our application's local variables, as well as reference information for our methods. Stacks use **LIFO** for efficient access to local variables and method calls. Limiting factors of stacks include their limited lifespan and size. This makes the use of stacks for long-term object storage impractical.

We push objects onto the heap when we use Java's new keyword. Heaps provide us with dynamic memory allocation. When we use a heap, our objects can exist outside the scope of the methods that created them. This necessitates the use of the JVM's garbage collector to remove items from the heap that are unreachable.

The following table summarizes the differences between stacks and heaps in Java:

| Characteristic | Stack | Heap |
| --- | --- | --- |
| Storage | Local variables | Objects and their methods and properties |
| Memory management | LIFO order | Managed by the JVM garbage collector |
| Size | Limited | Larger than stacks |
| Lifetime | Variables exist until the declaring method ends | Objects exist until they are no longer reachable |
| Performance | Fast allocation and deallocation | Slower than stacks |

Table 6.1 – Stack and heap comparison

The key differences between stacks and heaps are their scope of use, how the memory is managed, access speed, and storage size. In addition, there are risks of errors at runtime. For example, a stack can run out of memory, which leads to a StackOverflowError error. If a heap runs out of memory, an OutOfMemoryError error can be thrown. We need to handle those errors. Let's look at an example:

```
public static void main(String[] args) {
  // Catch StackOverflowError
  try {
    // your code goes here
  } catch (StackOverflowError e) {
    System.out.println("Caught StackOverflowError");
  }
  // Catch OutOfMemoryError
  try {
    // your code goes here
```

```
  } catch (OutOfMemoryError e) {
    System.out.println("Caught OutOfMemoryError");
  }
}
```

As you can see, we used `try-catch` blocks to trap the errors. Also, it is important to know that these errors are instances of `Error`, not `Exception`. So, these are indeed errors, not exceptions.

## Memory management with garbage collection

One of the most prized features of the Java programming language is its **garbage collection**. This automatic memory deallocation can take a lot of responsibility off the developer's shoulders, but there are some disadvantages as well. Let's take a deeper look at Java's seemingly straightforward garbage collection.

Java's garbage collector identifies objects that are no longer reachable in the application. Once identified, those objects are removed and the memory they used is deallocated, making it available to the application.

While we can applaud the garbage collector's efforts and appreciate the memory it frees up for our applications, there can be an impact on performance. When we have frequent garbage collection cycles, pauses and reduced responsiveness can be introduced at runtime. Mitigation strategies include minimizing object creation and implementing timely object disposal. These strategies can reduce the frequency of garbage collection.

We will look at how to implement these strategies later in this chapter.

## Optimization techniques

There are several strategies that we can adopt to help optimize memory usage in our Java applications:

- Limit object creation; only create them when absolutely needed
- Avoid creating objects within loops
- Use local variables and object pooling (covered in *Chapter 4*)
- Implement object immutability (covered later in the *Object immutability* section of this chapter)
- Use profiling tools for memory usage (see *Chapter 14*)

It is important to take a purposeful and informed approach to memory management in our Java applications. To do this, we need to understand memory limitations and how memory is allocated and deallocated.

## Object pooling

As you may recall from *Chapter 4*, object pooling is an important design pattern we can use to create a set of objects that can be kept in a pool, ready for use, instead of allocating them when we need them and deallocating them when they go out of scope. Object pooling helps us to efficiently manage system resources by reusing objects. The benefits are especially noticeable when our objects are large and take significant time to create.

## Initialization approaches

There is more than one way to create objects in Java and our initialization approach can significantly impact the overall performance of our Java applications and how our memory is used. Let's look at four approaches: direct, lazy, pooling, and builder.

### Direct initialization

The most common method that's used to create a new object is the **direct initialization** method. As shown in the following example, it's straightforward. We use constructors or initializers with this method:

```
Corgi myCorgi = new Corgi("Java", 3);
```

The benefits of this method are that it is easy to understand and program. It can also be used in many scenarios where objects need to be created. The disadvantages of this method include that it can lead to unnecessary object creation, which is the opposite of our goal to minimize object creation. This is especially evident when the direct initialization method is used inside loops or in methods that are called frequently. Another disadvantage is that the objects that are created using this method cannot be reused.

### Lazy initialization

The strategy of delaying an object's creation until it is needed by the application is referred to as **lazy initialization**. As illustrated in the following code snippet, the object isn't created unless a specific condition exists:

```
Corgi myCorgi = null; // Initialize to null
// ...
if (someCondition) {
    myCorgi = new Corgi("Java", 3); // Create the object when needed
}
```

An advantage of this method is that object creation is minimized until it is required. Additionally, memory usage is reduced when we conditionally create several objects. However, this strategy results in increased code complexity and can introduce synchronization issues when dealing with multi-threaded environments.

## Object pooling

We can also use object pooling to create a pool of pre-initialized objects that can be used multiple times, taking them from and returning them to the pool as needed. The following code snippet shows how that code would be structured:

```
ObjectPool<Corgi> corgiPool = new ObjectPool<>(Corgi::new, 10); //
Create a pool of 10 Corgi objects
// ...
Corgi myCorgi = corgiPool.acquire(); // Get obj from pool
// ...
corgiPool.release(myCorgi); // Release obj to the pool
```

You can revisit *Chapter 4* for greater detail on object pooling.

## Builder pattern

Another increasingly popular method of object creation is using the builder pattern. This is a design pattern that treats object construction and its representation separately. This method empowers us to create multi-attribute objects one step at a time. The following code snippet illustrates the build pattern concept:

```
CorgiBuilder builder = new CorgiBuilder();
builder.setName("Java");
builder.setAge(3);
Corgi myCorgi = builder.build();
```

One benefit of this method is that it introduces object construction flexibility. This can be very useful when you're creating complex objects. Another benefit is that it permits us to set selected attributes by having a complex constructor. The primary disadvantage of the builder pattern method is that it requires us to define a separate builder class for each type of object that's used in our applications. This can significantly increase our code's complexity and decrease its readability and maintainability.

# Overhead reduction

When we consider the goal of minimizing object creation, we should consider object pooling, object cloning, and object serialization. We touched on object pooling earlier in this chapter and provided in-depth coverage in *Chapter 4*. We will save our discussion on object cloning until later in this chapter. For now, know that it can be associated with overhead reduction.

The third concept is object serialization. Fortunately, Java allows us to convert objects to and from binary form. We typically use this for object persistence and can also use it to create copies of objects with reduced overhead.

Here's an example of how we can serialize (convert into binary) and deserialize (convert back to an object):

```
// Serialize the object
ByteArrayOutputStream outputStream = new ByteArrayOutputStream();
ObjectOutputStream objectOutputStream = new
ObjectOutputStream(outputStream);
objectOutputStream.writeObject(originalCorgi);

// Deserialize the object
ByteArrayInputStream inputStream = new
ByteArrayInputStream(outputStream.toByteArray());
ObjectInputStream objectInputStream = new
ObjectInputStream(inputStream);
Corgi clonedCorgi = (Corgi) objectInputStream.readObject();
```

Reducing overhead when creating objects should be a key consideration when our applications are expected to have high performance.

# Object immutability

Let's continue learning how to minimize object creation to increase the performance of our Java applications. **Object immutability** refers to an object that cannot be modified once it has been instantiated. Immutability can be considered a property or characteristic of an object. The advantages of making objects **immutable** include system predictability and performance.

Let's start with a brief overview of object immutability.

## Immutability overview

Object immutability is not a new concept, but it is an important one. The general premise is that we create an object with all the desired attributes and behaviors, and then prevent it from being **mutated** (changed) throughout the object's life cycle.

Immutable objects and considered safe because they cannot be changed. This means that we can share these objects in multi-threaded environments without the requirement for synchronization. So, concurrent programming is simplified. You will learn more about concurrency in *Chapter 9*.

Immutable objects are known to have an immutable state, be safe for sharing, have predictable behavior, and align with functional programming principles.

# Best practices

Creating immutable objects requires more than simply setting an attribute; it requires adherence to certain best practices. We want our immutable objects to be robust, the code to be maintainable, and positively contribute to the application's overall performance.

Understanding these best practices is key to proper implementation.

## *Declare as final*

The first best practice for creating immutable objects is to ensure all attributes are declared as `final`. The following code shows how to declare the class as `final`, as well as the two variables:

```
public final class ImmutableExample1 {
    private final int value;
    private final String text;
    public ImmutableExample1(int value, String text) {
        this.value = value;
        this.text = text;
    }
    // Getter methods...
}
```

Adhering to this best practice ensures the attributes cannot be changed once the object is created. This can also help us detect any attempts to change the object at runtime.

## *Complete constructor*

The second best practice is simply to ensure that all attributes of a class are initialized in the constructor. It is important to initialize all fields within the constructor. The goal is to ensure that the object is fully defined when it is created. Remember, we won't be able to make changes to the object later.

## *Avoid setters*

When you've been a Java developer for a considerable time, you'll likely create setters and getters in your classes automatically. We don't use a checklist; it just becomes a habit. In the case of immutable objects, we don't want to give our applications the ability to call a setter since no changes should be made to the object after it's created. The following code snippet shows how to create a standard class with a constructor. There is a getter method, but no setters:

```
public final class ImmutableExample2 {
    private final int value;
    private final String text;
```

```
    public ImmutableExample2(int value, String text) {
        this.value = value;
        this.text = text;
    }

    public int getValue() {
        return value;
    }

    public String getText() {
        return text;
    }
}
```

Since we shouldn't be calling any setters on immutable objects, it's important not to include them in our classes.

### Defensive copies

When we return a reference from an immutable object to an internal mutable object, it's important to return a **defensive copy**. This prevents any external modification. The following code snippet demonstrates how we should implement this best practice:

```
public final class ImmutableExample3 {
    private final List<String> data;

    public ImmutableExample3(List<String> data) {
        // Create a defensive copy to ensure the list cannot be
        // modified externally
        this.data = new ArrayList<>(data);
    }

    public List<String> getData() {
        // Return an unmodifiable view of the list to prevent
        // modifications
        return Collections.unmodifiableList(data);
    }
}
```

Using this approach helps ensure that the object's state remains immutable.

## *Annotation*

Our last best practice when it comes to creating immutable objects is to use the @Immutable annotation. Here, we are using the **Project Lombok** library:

```
import lombok.Immutable;
@Immutable
public final class ImmutableExample4 {
    private final int value;
    private final String text;

    public ImmutableExample4(int value, String text) {
        this.value = value;
        this.text = text;
    }
    // Getter methods...
}
```

When we use this annotation, we can benefit from auto-generated code making us more efficient. Note that this annotation may not be available in later versions of Lombok.

## Performance advantages

As you have learned so far, object immutability offers several benefits to us and our applications. One category of benefits is performance. Here's a list of performance advantages:

- **Predicable state**: When using immutable objects, we can rely on their state to remain constant throughout their lifespan.

- **Garbage collection efficiency**: Using immutable objects reduces how often object collection and disposal functions have to run.

- **Safe caching**: We can safely cache immutable objects and even share them among multiple threads without data corruption concerns.

- **Reduced overhead**: Because immutable objects are thread-safe, we don't need to use synchronization mechanisms in multi-thread environments.

- **Easier parallelization**: We can simplify concurrent programming and parallel programming when we use immutable objects.

- **Functional programming advantage**: As mentioned previously, immutable objects align with functional programming. In that programming paradigm, functions produce predictable results without side effects.

Understanding the performance advantages of using object immutability can encourage us to adopt this approach, which can result in sufficiently more performant Java applications.

## Custom classes

We previously reviewed best practices for implementing immutable objects. In addition to those best practices, we should implement the `equals` and `hashCode` methods. Let's look at that in code:

```java
@Override
public boolean equals(Object o) {
    if (this == o) return true;
    if (o == null || getClass() != o.getClass()) return false;
    CustomImmutable that = (CustomImmutable) o;
    return value == that.value && Objects.equals(text, that.text);
}
@Override
public int hashCode() {
    return Objects.hash(value, text);
}
```

As you can see, when we want to perform equality testing and ensure compatibility with certain data structures, we can override the `equals` and `hashCode` methods in our custom immutable classes. When we do this, we must ensure we consider all attributes that can contribute to equality.

## String classes

As you likely know, strings are a commonly used data type and they are, by their very nature, immutable. Let's look at how strings work as immutable objects so that we can better understand how to design immutable objects.

Strings are indeed immutable and even when we think we are modifying them, Java is creating a new object. As an example, consider the following lines of code:

```java
String original = "Java";
String modified = original.concat(", is the name of my Corgi");
```

As you can see, when we call the `concat` method on the first string, a new string object is created.

String immutability provides us with several advantages, such as thread safety, predictable behavior, and efficiency with string manipulations. Under the hood, Java maintains a string pool, also referred to as an intern pool, to store unique string literals. This is another advantage of string immutability. Let's look at this in code:

```java
String s1 = "Java"; // Stored in the string pool
String s2 = "Java"; // Reuses the same string from the pool
```

A fifth advantage of string immutability is security. This means we can confidently use strings for sensitive data, such as banking information, passwords, and cryptographic keys, because unintentional modification is prevented.

# Garbage collection

We have already established that memory management is important when our goal is to have our Java applications perform at a high level. We also looked at how garbage completion works and what its benefits are.

## Garbage collection implications

The automatic nature of Java's garbage collection results in many developers ignoring it. They take garbage collection for granted and do not implement any best practices. This is okay for small projects that are not data- or memory-intensive. Let's look at two ways that garbage collection can impact our applications and memory management:

- **Application pauses**: Frequent garbage collection cycles can result in our application pausing. The type of garbage collection and heap size are key determiners for the length of these pauses.
- **Memory overhead**: Garbage collection increases memory overhead. CPU cycles and memory resources are impacted each time garbage collection runs.

There are several approaches we can take to help mitigate the impact garbage collection has on our applications:

- Take a purposeful approach to managing the life cycle of our objects
- Avoid unnecessary object creation
- Reuse objects
- Implement object pooling
- Use immutable objects

In the next section, we will look at how object cloning relates to garbage collection.

## Object cloning

As you might assume, **object cloning** is when we create a new object that is a duplicate of an existing one. Its relevance to garbage collection is due to its potential impact on how objects are managed and disposed of. The type of impact is influenced by the type of cloning used. Let's look at two types of cloning: shallow and deep.

## Shallow cloning

The process of **shallow cloning** involves creating a new object by copying the contents of the original object. It's important to note that if the original object contains any references to other objects, the clone will point to the same objects. This is an expected behavior; we are cloning all the objects so that they include references, but not the referenced objects. Let's look at a brief example in code:

```java
class Corgi implements Cloneable {
    private String name;
    private String authenticityCertificate;

    // Constructor and getters ...
    @Override
    public Object clone() throws CloneNotSupportedException {
        return super.clone();
    }
}
```

As you can see, when we clone a `Corgi` object, the new Corgi will share the same `authenticityCertificate` as the original. If that field is mutable, then changes that are made to it through one reference will take effect in both the original and cloned Corgi objects.

## Deep cloning

When we create a **deep clone**, we still create a new object, but it also recursively copies all objects that are referenced by the original object. This method of cloning ensures that the new object and its sub-objects are independent of the original object. Let's look at this in code:

```java
class Corgi implements Cloneable {
    private String name;
    private Address address;

    // Constructor and getters...

    @Override
    public Object clone() throws CloneNotSupportedException{
        Corgi clonedCorgi = (Corgi) super.clone();
        clonedCorgi.address = (Address) address.clone(); // Deep copy of
the Address object
        return clonedCorgi;
    }
}
```

As you can see, when we clone a `Corgi` object, it is a new `Corgi` object and a new `Address` object is created. With this method, we can make changes to one object without impacting the other.

# Design patterns

Design patterns are time-tested solutions to common software problems. They can be considered a set of best practices and are widely used for Java development. Concerning strategic object creation and immutability, two design patterns deserve our attention:

- Singleton pattern
- Factory pattern

> **What a design pattern is not**
>
> Design patterns are structured approaches to solving known problems. They are not algorithms, templates, libraries, or even code snippets. Instead, they offer high-level guidance that we can follow.

Let's look at each of these patterns so that we understand how their use can help improve the performance of our Java applications.

## Singleton pattern

The **singleton design pattern** ensures that there is only one instance of a class and then provides global access to that instance. This pattern is often used when an application is managing database connections, resource management, logging, caching, or configuration settings.

Let's look at a simple implementation approach for this pattern:

```java
public class Singleton {
    private static Singleton instance;

    private Singleton() {
        // Private constructor to prevent instantiation
    }

    public static Singleton getInstance() {
        if (instance == null) {
            instance = new Singleton();
        }
        return instance;
    }
}
```

The preceding code is a standard example. As you can see, this class prohibits more than one instance from being created. Additionally, the `getInstance()` method is how we provide global access to the instance of the `Singleton` class.

# Factory pattern

The **factory design pattern** involves a superclass and an interface for creating objects in it. This pattern permits subclasses to alter what can be created. It promotes loose coupling between the superclass and the subclasses being created. The most common components of this pattern are as follows:

- **Abstract factory**: This is an interface that declares the method that's used to create objects.
- **Concrete factory**: This is a class that implements the abstract factory interface. It creates concrete objects.
- **Product**: This is the object that's created by the factory.
- **Concrete product**: This is the class that implements the product interface.

The advantages of using this pattern include separation of concerns, code reusability, flexibility, and encapsulation. Next, we will look at several code snippets that illustrate simple implementation examples.

First, here is an example of an abstract product:

```
interface Product {
    void create();
}
```

Now, let's look at an example of concrete products:

```
class ConcreteProductA implements Product {
    @Override
    public void create() {
        System.out.println("Creating Concrete Product A");
    }
}
class ConcreteProductB implements Product {
    @Override
    public void create() {
        System.out.println("Creating Concrete Product B");
    }
}
```

The following code snippet illustrates how to implement an abstract factory:

```
interface Factory {
    Product createProduct();
}
```

Lastly, the following code demonstrates how to implement concrete factories:

```
class ConcreteFactoryA implements Factory {
    @Override
    public Product createProduct() {
        return new ConcreteProductA();
    }
}
  class ConcreteFactoryB implements Factory {
    @Override
    public Product createProduct() {
        return new ConcreteProductB();
    }
}
```

The factory pattern can be a valuable tool when you need flexibility when it comes to creating objects based on specific requirements or conditions.

## Summary

This chapter focused on strategic object creation and immutability, two closely related and equally important topics. We looked at various aspects of these topics to help improve the performance of our Java applications. Specifically, we looked at minimizing object creation, object immutability, garbage collection, and design patterns. You should now have a better understanding and deep appreciation for the importance of strategic object creation, as well as its best practices and implementation strategies. You should also have a firm grasp of the concept of object immutability.

In the next chapter, *String Objects*, we will take a deeper look into string objects while covering topics such as proper string pooling, lazy initialization, and string operation strategies.

# 7
# String Objects

In an effort to analyze every aspect of our Java applications to ensure that they perform at a highly efficient rate, we need to consider string objects. Strings are a big part of Java applications and are used for limitless purposes from a simple list of names to complex data storage needs, such as with a bank's databases. The creation, manipulation, and management of these objects should be a primary concern.

This chapter focuses on the efficient use of string objects in our Java applications. The first concept is proper string pooling. We will examine this concept and explore best practices for using string pooling for high performance. The chapter also introduces the concept of lazy initialization, examines its benefits, and illustrates implementation with sample code. Lastly, we will look at additional string operation strategies, to include advanced string manipulation techniques.

In this chapter, we will cover the following main topics:

- Proper string pooling
- Lazy initialization
- Additional string operation strategies

## Technical requirements

To follow the examples and instructions in this chapter, you will need the ability to load, edit, and run Java code. If you have not set up your development environment, please refer back to *Chapter 1*.

The finished code for this chapter can be found here: https://github.com/PacktPublishing/ High-Performance-with-Java/tree/main/Chapter07.

# Proper string pooling

Our overarching concern is to ensure that our Java applications perform at a high level. To that end, memory management is a critical concern. It is important that we design, test, and implement techniques for optimizing memory usage. **String pooling** is one such technique, whose focus is to enable the reuse of string objects for greater application efficiency. The efficiency gains stem from reducing memory overhead.

String pooling is an important concept that is anchored to the sharing of string values as an alternative to creating new string instances. If your application uses string literals frequently, then you should find string pooling especially useful.

> **String literal**
>
> A string literal is a fixed value bookended by double quotes. For example, the `"Read more books."` component of the following statement is a string literal:
>
> `System.out.println("Read more books.");`

To examine string pooling, we will start with the concept of **string interning**, then review best practices, and complete our discovery with code examples using Java. To take this concept one step further, we will look at string pooling for database queries.

## String interning

String interning is an interesting concept. It is a memory reduction technique that allows more than one string to use the same memory location. As you might expect, the contents of those strings must be identical for this to work. Memory reduction is possible because we can eliminate duplicate objects. String interning relies on string pooling for this to work. As we covered in *Chapter 4*, pooling uses a special heap area to store the string literals.

Let's try to think like the **Java Virtual Machine** (**JVM**) does. When a string literal is encountered by the JVM, it first checks to see whether it is in the string pool. If it is not found, then the JVM creates the new string object and stores it on the heap. If the string literal is already on the heap, the JVM simply returns a reference to the existing object. In Java, we use the `intern()` method, part of the `String` class, to accomplish this. We will look at an example later in the *Code examples* section of this chapter.

## Best practices

In general, it is considered a good idea to employ string interning in our applications. When implementing string interning, there are two important considerations to be aware of. First, we should not overuse string interning. The performance and memory usage benefits are clear, but if our applications use interning on very large strings, then we might consume more memory than we want to.

Another best practice to consider is avoiding **implicit interning**. As we learned from our JVM discussion earlier, string literals are automatically interned by the JVM when they are hardcoded. Dynamically created strings, on the other hand, are usually not automatically interned. When we want to be explicit in our interning, we need to use the `intern()` method. This way, we can ensure consistent behavior in our application.

## Code examples

Let's look at a simple example to demonstrate string interning. The following code starts by defining two string literals, `s1` and `s2`. Both strings have the same values. A third string, `s3`, is created using the `intern()` method. The final two statements are used to compare references:

```
public class CorgiStringIntern {
    public static void main(String[] args) {
        String s1 = "corgi";
        String s2 = "corgi";

        String s3 = new String("corgi").intern();
        System.out.println("Are s1 and s2 the same object? " + (s1 ==
        s2));
        System.out.println("Are s1 and s3 the same object? " + (s1 ==
        s3));
    }
}
```

As you can see from our application's output, both reference comparisons are evaluated as `true`:

```
Are s1 and s2 the same object? true
Are s1 and s3 the same object? true
```

### String pooling for database queries

SQL and other database queries are often created dynamically in our code and can become quite complex. As with other large strings, we can use string pooling to help ensure that our application does not unnecessarily create string objects. Our goal is to reuse any commonly used database queries instead of recreating them multiple times at runtime.

# Lazy initialization

The decision to not instantiate an object until it is needed is called **lazy initialization**. The word *lazy* normally has a negative connotation; however, in the context of software development, lazy initialization is a performance optimization approach with the goal of managing an application's overhead. This design pattern is especially useful when detailing very large objects.

> **Should I implement lazy initialization?**
>
> If you have string objects that are large or complex and require significant overhead at initialization time, then you should consider lazy initialization.

Lazy initialization does not reduce the number of string objects created; rather, it delays the initialization until it is required by your application for processing. This delaying technique can help you with memory optimization. Let us look at a narrative example before we examine the source code.

Imagine you have a legacy application that uses very sophisticated string objects. When your application opens, a loading screen is displayed and, behind the scenes, the application creates several string objects to support the application's normal operations. Users have complained that the application "takes forever to load" and that the computer often locks up when trying to launch the application. After reviewing the source code, you realize that many of the string objects created when the application is launched are only used by the application when the user selects certain functions. Your solution is to implement lazy initialization, so these string objects are only created when and if they are needed.

## Code examples

Now, let us look at how we can implement the lazy initialization design pattern in Java. We will start by importing `java.util.function.Supplier`. It is an interface that provides a single `get()` method, which we will use to retrieve a value:

```
import java.util.function.Supplier;
```

Next, we declare our class and use `StringBuilder` to generate a complex string using the `append()` method multiple times:

```
public class LazyInitializationExample {
  private Supplier<String> lazyString = () -> {

    StringBuilder stringBuilder = new StringBuilder();
    stringBuilder.append("Java is an incredibly ");
    stringBuilder.append("flexible programming language ");
    stringBuilder.append("that is used in a wide range of
    applications. ");
    stringBuilder.append("This is a complex string ");
    stringBuilder.append("that was lazily initialized.");
    return stringBuilder.toString();
};
```

Staying inside the `LazyInitializationExample` class, we make a call to the `getLazyString()` method. This method is used to create or retrieve the complex string only when it is needed – not before:

```
public String getLazyString() {
  return lazyString.get();
}
```

The last part of our code is the `main()` method. When this method is run, we access the lazily initialized string:

```
public static void main(String[] args) {
  LazyInitializationExample example = new
  LazyInitializationExample();

  String lazyString = example.getLazyString();
  System.out.println("Lazily initialized string: " + lazyString);
  }
}
```

The output of the program is as follows:

```
Lazy-initialized string: Java is an incredibly flexible programming
language that is used in a wide range of applications. This is a
complex string that was lazily initialized.
```

## Best practices

The concept of lazy initialization is straightforward and the decision to implement it is easy. If you have large string objects, then you should try this approach. There are two best practices to consider with lazy initialization. First, we should implement synchronization when we are working in a multiple-thread environment. This is to help promote thread safety. Another best practice is to avoid overuse. When we overuse lazy initialization, our code might become difficult to read and maintain.

# Additional string operation strategies

String pooling and lazy initialization are excellent optimization strategies that can help improve the overall performance of our Java applications. In addition to these strategies, we can ensure that our string concatenation operations are efficient, that we properly leverage regular expressions, and that we efficiently handle large text files. This section reviews techniques in each of those areas.

## Concatenation

String concatenation – the joining of two or more strings into one – using the plus (+) operator often results in inefficient code. This concatenation creates a new string object, which we want to avoid. Let's look at two alternatives that offer better performance.

This first alternative uses `StringBuilder`. In the following example, we create a `StringBuilder` object, append five string literals to it, convert the `StringBuilder` object to `String`, and then output the results:

```java
public class ConcatenationAlternativeOne {
  public static void main(String[] args) {
    StringBuilder myStringBuilder = new StringBuilder();

    myStringBuilder.append("Java");
    myStringBuilder.append(" is");
    myStringBuilder.append(" my");
    myStringBuilder.append(" dog's");
    myStringBuilder.append(" name.");

    String result = myStringBuilder.toString();

    System.out.println(result);
  }
}
```

Another alternative is to use `StringBuffer`. The following program is similar to our `StringBuilder` example but uses `StringBuffer` instead. As you can see, both methods are implemented in the same manner:

```java
public class ConcatenationAlternativeTwo {
  public static void main(String[] args) {
    StringBuffer myStringBuffer = new StringBuffer();

    myStringBuffer.append("Java");
    myStringBuffer.append(" is");
    myStringBuffer.append(" my");
    myStringBuffer.append(" dog's");
    myStringBuffer.append(" name.");

    String result = myStringBuffer.toString();

    System.out.println(result);
  }
}
```

The difference between the `StringBuilder` and `StringBuffer` alternatives to string concatenation is that `StringBuffer` offers us thread safety, so it should be used in multiple-thread environments; otherwise, `StringBuilder` is an alternative to using the plus (+) operator for string concatenation.

# Regular expressions

**Regular expressions** provide us with an excellent method of pattern matching as well as string manipulation. Because these expressions can be processor- and memory-intensive, it is important to learn how to use them efficiently. We can optimize the use of regular expressions by compiling our patterns only once and reusing them as needed.

Let's look at an example. In the first part of our application, we import the `Matcher` and `Pattern` classes:

```
import java.util.regex.Matcher;
import java.util.regex.Pattern;
```

Next, we establish our email pattern regex:

```
public class GoodRegExExample {
  public static void main(String[] args) {
    String emailRegex = "^[a-zA-Z0-9._%+-]+@[a-zA-Z0-9.-]+\\.[a-zA-Z]
    {2,}$";
```

This next section of code creates a list of sample email addresses. We will later check these for validity:

```
String[] emails = {
  "java@example.com",
  "muzz.acruise@gmail.com",
  "invalid-email",
  "bougie@.com",
  "@example.com",
  "brenda@domain.",
  "edward@domain.co",
  "user@example"
};
```

The next statement compiles the regex pattern:

```
Pattern pattern = Pattern.compile(emailRegex);
```

This last section of code iterates through each email address in our list:

```
    for (String email : emails) {
      Matcher matcher = pattern.matcher(email);
      if (matcher.matches()) {
        System.out.println(email + " is a valid email address.");
      } else {
        System.out.println(email + " is an invalid email address.");
      }
    }
  }
}
```

This simple implementation of regular expression checks for email address formatting and demonstrates how to compile the regex pattern once and use it elsewhere in the application as an efficient alternative to compiling the pattern each time we need to perform the email address validation operation.

## Large text files

Some Java applications can process very large text files, even book-length files. It is not advisable to load the file's complete content at one time. Our applications can quickly run out of available memory and cause undesirable runtime results. An alternative is to use a buffered approach to read the text files in segments.

The following example assumes that there is a text file in the local directory. We read each line using `BufferedReader`:

```java
import java.io.BufferedReader;
import java.io.FileReader;
import java.io.IOException;

public class LargeTextFileHandlingExample {
  public static void main(String[] args) {
    String filePath = "advanced_guide_to_java.txt";

    try (BufferedReader reader = new BufferedReader(new
    FileReader(filePath))) {
      String line;
      int lineCount = 0;

      while ((line = reader.readLine()) != null) {
        System.out.println("Line " + (++lineCount) + ": " + line);
      }
    } catch (IOException e) {
      e.printStackTrace();
    }
  }
}
```

This approach can help us manage our memory and improve the overall performance of our applications.

# Summary

This chapter focused on how to create, manipulate, and manage strings to contribute to the overall performance of your Java applications. You should now understand string pooling and have the confidence to use it effectively based on best practices. Lazy initialization should now be a strategy you consider implementing in your future applications when dealing with the extensive use of strings and you are concerned about thread safety. The chapter also introduced advanced string operation strategies to help give you choices when designing your Java applications for high performance.

In the next chapter, *Memory Leaks*, we will look at what memory leaks are, how they are created, and what effects they have on our applications. We will look at strategies for avoiding memory leaks to improve the performance of our Java applications.

# 8

# Memory Leaks

Memory leaks occur as a result of improper memory management and can directly impact the performance of an application. These leaks occur when memory is improperly deallocated or when it is allocated but becomes inaccessible. Improper memory management not only negatively impacts performance but can also hinder scalability, result in system crashes due to `OutOfMemoryError`, and ruin the user experience. Many developers implicitly trust Java's garbage collector (covered in *Chapter 1*) to manage memory while their applications run; however, despite the garbage collector's incredible capabilities, memory leaks are a persistent issue.

The garbage collector is not faulty; rather, memory leaks occur when the garbage collector is unable to reclaim memory that stores objects that are no longer needed by the application. Improper referencing is the primary culprit, and fortunately, we can avoid this. This chapter provides techniques, design patterns, coding examples, and best practices to avoid memory leaks.

This chapter covers the following main topics:

- Proper referencing
- Listeners and loaders
- Caching and threads

By the end of this chapter, you should have a comprehensive understanding of what can lead to memory leaks at runtime and the potential devastation they can inflict on our Java applications, and you will know how to prevent them purposefully and efficiently. You can gain confidence in implementing your own memory leak prevention strategy by experimenting with the sample code provided.

## Technical requirements

To follow the examples and instructions in this chapter, you will need the ability to load, edit, and run Java code. If you have not set up your development environment, refer to *Chapter 1*.

The finished code for this chapter can be found here: `https://github.com/PacktPublishing/High-Performance-with-Java/tree/main/Chapter08`.

# Proper referencing

It is undeniable – memory leaks can result in a gradual decrease in resource availability and lead to sluggish systems and potential system crashes. The good news is that there are two major components that offer a solution. One component is the garbage collector that is part of the **Java Virtual Machine (JVM)**. It is highly capable and one of the shining characteristics of the Java language. The second, and more important, component is the developer. As Java developers, we have the power to minimize and even eliminate memory leaks by taking a purposeful approach to memory management in our code.

To support the developer component of the solution to eradicate memory leaks, this section will focus on how to properly reference objects so that they do not lead to memory leaks, how to identify memory leaks, and the strategies to avoid them.

## An introduction to references

Perhaps the most important aspect of avoiding memory leaks is to use proper referencing. This should be encouraged, since it puts control in the developer's hands. Java offers a great toolbox to aid our efforts in this area. Specifically, there are several types of references, each with its own purpose and associated garbage collector behavior.

> **References**
>
> In Java, references are pointers to memory locations and are a critical component of memory management and memory leak mitigation.

Let's examine the various types of referencing in Java, strong, soft, weak, and phantom, so we can determine which method is the most appropriate for any given use case. Remember, our overall purpose is to have efficient memory management and avoid memory links to increase the overall performance of our Java applications.

### Strong references

**Strong referencing** is the most important reference type for us to focus on. Not only is it the default reference type in Java, but it is also the most common source of memory leaks. Objects that have a strong reference type are not eligible for garbage collection. To create a strong reference, we simply use a variable to directly reference an object.

In the following example code, we create a strong reference to an object using the `sampleCorgiObject` variable. As long as that variable contains a reference to the `SampleCorgiObject` instance, we will have that object in memory and the garbage collector will not be able to deallocate its memory:

```
public class CH8StrongReferenceExample {
  public static void main(String[] args) {
    SampleCorgiObject sampleCorgiObject = new SampleCorgiObject();
```

```
      System.out.println(sampleCorgiObject);
  }
  static class SampleCorgiObject {
    @Override
    public String toString() {
      return "This is a SampleCorigObject instance.";
    }
  }
}
```

This is the default reference type in Java for a reason. The typical use case is when we have objects that we need the entire time our application is running, such as configuration properties. We should use strong referencing with caution, especially when the objects are large. A best practice is to set our references to null as soon as they are no longer needed. This will empower the garbage collector to deallocate associated memory.

## Soft references

**Soft references** present a flexible memory management approach for high memory efficiency. This approach permits the JVM to collect the referred objects when the system runs low on available memory. The goal of soft references is to avoid OutOfMemoryError and system crashes. When we create soft references, those objects are retained in memory only when there is sufficient space. This makes soft referencing an ideal solution to cache large objects.

It is important to understand that the JVM collects these objects that have soft references to them only when necessary. The JVM's garbage collector will collect everything else it can first and then, as a final effort, collect objects with soft references, and only then if the system is running out of memory.

To implement a soft reference, we use the SoftReference class that is a part of the java.lang-ref package. Let's look at an example in code.

We start our application by importing SoftReference. Then, we create our class header and initially create MyBougieObject, using strong referencing by myBougieObject. We then wrap it with SoftReference<MyBougieObject> to establish the soft reference. Note that we set myBougieObject to null, which ensures that our MyBougieObject instance is only accessible through the soft reference we created:

```
import java.lang.ref.SoftReference;

public class CH8SoftReferenceExample {
  public static void main(String[] args) {
    MyBougieObject myBougieObject = new MyBougieObject("Cached
    Object");
```

```
SoftReference<MyBougieObject> softReference = new
SoftReference<>(myBougieObject);

myBougieObject = null;
```

In the next section of code, we attempt to review myBougieObject from the soft reference. We use System.gc() to provide a way to observe the behavior of our software references under normal conditions and then simulate memory pressure:

```
MyBougieObject retrievedObject = softReference.get();
if (retrievedObject != null) {
    System.out.println("Object retrieved from soft reference: "
        + retrievedObject);
} else {
  System.out.println("The Object has been garbage collected by the
  JVM.");
}
System.gc();
retrievedObject = softReference.get();
if (retrievedObject != null) {
  System.out.println("Object is still available after requesting
  GC: " + retrievedObject);
} else {
  System.out.println("The Object has been garbage collected after
  requesting GC.");
}
  }
 }
}
```

This last section implies our MyBougieObject class:

```
class MyBougieObject {
  private String name;

  public MyBougieObject(String name) {
    this.name = name;
  }
  @Override
  public String toString() {
    return name;
  }
}
```

Here is what the output might look like. Of course, the results will depend on your system's available memory:

```
Object retrieved from soft reference: Cached Object
Object is still available after requesting GC: Cached Object
```

### Weak references

So far, we have covered **strong references** that prevent garbage collection and **soft references** that allow garbage collection as a last-ditch effort to reclaim memory. **Weak references** are unique in that they permit garbage collection only if the weak reference is the only reference to the specific object. This approach can be especially useful when we want more flexibility in our memory management solutions.

A common use of weak references is in caching when we want objects to remain in memory but not prevent them from being reclaimed by JVM's garbage collection when a system runs low on memory. To implement a weak reference, we use the `WeakReference` class that is a part of the `java.lang-ref` package. Let's look at an example in code. We start our code with the necessary `import` statement and the class declaration. As you can see in the following code block, we wrap our `CacheCorgiObject` in `WeakReference`, which can initially be accessed via the weak reference. When we set the strong reference (`cacheCorgiObject`) to null, we call `System.gc()` to invoke the JVM's garbage collector. Depending on your system's memory, the object might be collected, as it is available. Following the garbage collection, we call `weakCacheCorgiObject.get()`, and `null` is returned if the object collection took place:

```
import java.lang.ref.WeakReference;
public class CH8WeakReferenceExample {
  public static void main(String[] args) {
    CacheCorgiObject cacheCorgiObject = new CacheCorgiObject();
    WeakReference<CacheCorgiObject> weakCacheCorgiObject = new
    WeakReference<>(cacheCorgiObject);
    System.out.println("Cache corgi object before GC: " +
    weakCacheCorgiObject.get());

    cacheCorgiObject = null;
    System.gc();
    System.out.println("Cache corgi object after GC: " +
    weakCacheCorgiObject.get());
  }
}
class CacheCorgiObject {
@Override
  protected void finalize() {
    System.out.println("CacheCorgiObject is being garbage collected");
  }
}
```

Here is a sample output from our program. Results will vary, depending on the system's available memory:

```
Cache corgi object before GC: CacheCorgiObject@7344699f
Cache corgi object after GC: null
CacheCorgiObject is being garbage collected
```

### Phantom references

Our last type of reference is a **phantom reference**. This reference type does not permit direct retrieval of the referenced object; instead, it provides a method to determine whether the object has been finalized and reclaimed by the JVM garbage collector. This happens without preventing the object from being collected.

Implementation requires two classes from the java.lang.ref package – PhantomReference and ReferenceQueue. Our example code demonstrates the garbage collector determining whether an object with a phantom reference is reachable. This means that the object has been **finalized** and is ready for garbage collection. When this is the case, the reference is queued, and our application is able to respond accordingly:

```java
import java.lang.ref.PhantomReference;
import java.lang.ref.ReferenceQueue;
public class CH8PhantomReferenceExample {
  public static void main(String[] args) throws InterruptedException {
    ReferenceQueue<VeryImportantResource> queue = new
    ReferenceQueue<>();

    VeryImportantResource resource = new VeryImportantResource();
    PhantomReference<VeryImportantResource> phantomResource = new
    PhantomReference<>(resource, queue);
    resource = null;
    System.gc();

    while (true) {
      if (queue.poll() != null) {
        System.out.println("ImportantResource has been garbage
        collected and its phantom reference is enqueued");
        break;
      }
    }
  }
}
class VeryImportantResource {
  @Override
  protected void finalize() throws Throwable {
    System.out.println("Finalizing VeryImportantResource");
```

```
        super.finalize();
    }
}
```

Adopting the phantom reference approach enables us to invoke our own memory-cleaning actions, based on an object with a phantom reference's collection. This approach does not interfere with the normal garbage collection operations.

---

**Using the finalize() method**

The `finalize()` method has been depreciated and is scheduled for removal from Java in a future release. It is used in the previous code example simply to demonstrate the phantom reference approach and does not suggest using `finalize()`. Consult the Java documentation for additional information.

---

Now that we have reviewed proper referencing, we should have an appreciation for how much developer intervention can solve the memory leak problem. Further, examination of the coding examples should lead to confidence in implementing proper referencing throughout our Java applications.

## Memory leak identification

One of the most important things we should implement in our Java applications is the ability to identify memory leaks. Our goal is to create high-performance Java applications, and memory leaks are counter to that goal. In order to identify memory leaks, we need to understand what symptoms indicate that a memory leak exists. These symptoms include increased garbage collection activity, `OutOfMemoryError`, and progressive performance degradation. Let's look at five methods to identify memory leaks.

One of the most important methods to identify memory leaks is to review our code to ensure that we follow best practices and avoid common pitfalls, such as failing to clear static collections, not removing **listeners** after use, and caches that grow out of control. We will cover listeners later in this chapter:

- A second method involves using a tool that can reveal which of our application's objects consumes the most memory.

- A third method is to use a tool to generate a heap dump, which is a moment-in-time snapshot of all objects currently in memory. This can help you analyze and detect potential issues.

When reviewing object retention, look for unusual or unwanted patterns. Examples include when objects have a longer life cycle than expected or a large number of objects of a specific type that you do not expect to see large numbers of.

A fifth method to identify memory leaks is to continually and iteratively test and profile your application. Once you identify a memory leak, you implement a fix, and should then retest.

We can use tools to help us identify memory leaks, including **JProfiler**, **YourKit**, **Java Flight Recorder** (**JFR**), **VisualVM**, and the Eclipse **Memory Analyzer Tool** (**MAT**). If you want to take memory leak identification seriously, you should research these tools to see how you can leverage their capabilities.

### Memory leak avoidance strategies

Proper object referencing, as detailed earlier in this chapter, is a primary strategy to avoid memory leaks in your Java applications. Identifying and fixing memory leaks is another strategy, albeit a reactive one. These two strategies are important and have already been covered. Let's briefly look at some additional strategies.

A third memory leak avoidance strategy is to properly manage collection objects. It is not uncommon for objects to be put into collections and then ignored or forgotten. This can result in memory leaks. So, to avoid this, we should develop our applications so that they regularly remove objects that are no longer required by our application. Using weak references can help with this. We should also be careful when using static collections. This type of collection has its life cycle linked with the class loader.

We should also be mindful of how we implement caches. The use of caches can significantly improve an application's performance but can also result in memory leaks. When implementing caches, we should use soft references, set finite cache size limits, and continuously monitor cache usage.

A fifth strategy is to continuously use a profiling tool and test your application. This strategy requires a never-ending dedication to detecting and removing memory leaks in your application. It is an important strategy that should not be taken lightly.

When we implement a set of memory leak avoidance strategies, we have a better chance of ensuring our applications have high performance.

Next, we will review how listeners and loaders can be used to help avoid memory leaks.

# Listeners and loaders

Several aspects of our Java applications can impact performance and, more specifically, result in memory leaks. Two of those aspects are **listeners** and **loaders**. This section looks specifically at **event listeners** and **class loaders** and includes strategies to mitigate the risks of using them, without sacrificing the power and efficiencies they can provide to our Java applications.

### Event listeners

Event listeners are used to allow objects to react to events. This approach is used heavily in interactive applications and is a critical component of event-driven programming. These listeners can result in a highly interactive application; however, if not properly managed, they can be the source of memory leaks.

To understand the issue better, it is important to note that event listeners must subscribe to event sources so that they can receive notifications that need to be acted upon (e.g., a button click, or a non-player character in a game entering a predefined zone). As you would expect, each event listener maintains references to the event sources they subscribe to. A problem arises when an event source is no longer needed but is referenced by one or more listeners; this prevents the garbage collector from collecting the event source.

Here are some best practices to avoid memory leaks when working with event listeners:

- Use weak references, as detailed earlier in this chapter.

- Explicitly deregister listeners from event sources when applicable.

- Implement static nested classes for listeners because they do not have implicit references to outer class instances. This approach should be used instead of implementing non-static inner classes.

- Align your event listener life cycles with those of their associated event sources.

## Class loaders

Class loaders enable us to dynamically load classes, making them a key component of **Java's Runtime Environment (JRE)**. Class loaders offer great power and flexibility through the support of polymorphism and extensibility. When our applications load classes dynamically, it illustrates that Java does not need to know about these classes at compile time. With this powerful flexibility comes the potential for memory leaks, which we need to mitigate, if not eliminate.

The JVM has a class loading delegation model that involves several class loader types:

- A **bootstrap class loader** that loads Java's core API classes.

- An **extension class loader** that loads classes from directories, as specified in the `java.ext.dirs` property.

- An **application class loader** that loads classes defined by the application. These are loaded from `classpath`.

Class loaders are necessary, and when the loaded classes are retained in memory for longer than necessary, they can introduce memory leaks. The culprits here are typically with static fields in classes holding reference to objects that should otherwise be collected by the garbage collector, objects of loaded classes that are referenced by objects that have long lifespans, and a cache that retains class instances without proper management. Here are a few strategies to mitigate these risks:

- Use weak references

- Minimize the use of static fields

- Make sure that custom class loaders are available to the garbage collection when no longer needed

- Monitor, profile, and refine as needed

When we have a thorough understanding of loaders and listeners as well as strategies to mitigate their associated risks, we improve our chances of minimizing memory leaks in our Java applications.

# Caching and threads

This section explores caching strategies, thread management, and the effective use of Java concurrency utilities. These are important concepts to embrace as we continue our journey of developing high-performance Java applications. We will explore these concepts, the associated best practices, and techniques to mitigate the risk of memory leaks introduced by their use.

## Caching strategies

In programming, we use **caching** to temporarily store data in memory locations to permit rapid access. This allows us to repeatedly access the data without causing lag or system crashes. The benefits of caching include more responsible applications and less load on longer-term storage solutions, such as databases and database servers. Of course, there are pitfalls. If we do not properly manage our caches, we can introduce significant memory leaks into our applications.

There are multiple caching strategies for us to consider; two of them you should already be familiar with, as they were covered earlier in this chapter, albeit not specific to caching. The first familiar strategy is to use weak references. When we use weak references with caching, we allow garbage collection when memory runs low. The second familiar strategy is using soft references. This strategy enables higher retention priority during garbage collection cycles.

Another caching strategy is referred to as **Least Recently Used** (**LRU**), and as the name suggests, we remove the least accessed items first, as they are the least likely to be used by our application again.

**Time to Live** (**TTL**) is another useful caching strategy. The TTL approach tracks cache insertion times and automatically expires items based on a prescribed amount of time.

One additional caching strategy is using **size-based eviction**. This strategy ensures that caches do not exceed the amount of memory you set as the maximum boundary. The boundary can be set in terms of memory usage or the total number of items.

When we implement caching, we should be mindful of introducing memory leaks due to poor implementation. This purposeful approach requires us to conduct capacity planning, establish an eviction policy for the LRU and TTL approaches, and monitor our system. This monitoring requires subsequent fine-tuning and retesting.

# Thread management

We use **threads** in our applications to facilitate multiple concurrent operations. This makes better use of modern CPUs and improves the responsiveness of our Java applications. When we manage our threads manually, we can introduce memory leaks due to the complexity and error-prone nature of managing threads.

We can create threads in Java by extending the `Thread` class, or even by implementing the `Runnable` interface. These methods are a direct and easy way to use threads; however, it is not a recommended approach for large systems because they directly create and utilize threads, resulting in significant overhead. Instead of the direct approach, consider using Java's **Executor** framework to abstract thread management from the main application.

Some best practices for thread management include the following:

- Favor Executor over direct thread creation
- Limit the number of thread pools you use to a minimum
- Support thread interruptions in your code
- Monitor and refine consistently

Let's look at a simple example to demonstrate proper thread usage. We will use the `Executor` framework. As you will see, the following application creates a fixed thread pool to run a set of tasks. Each task simulates a real-world operation by sleeping for a defined amount of time. As you walk through the code, you will see that the `Executor` framework is efficiently used to manage threads:

```java
import java.util.concurrent.ExecutorService;
import java.util.concurrent.Executors;
import java.util.concurrent.TimeUnit;

public class CH8ThreadManagementExample {
  public static void main(String[] args) {
    ExecutorService executor = Executors.newFixedThreadPool(5);
    for (int i = 0; i < 10; i++) {
      final int taskId = i;
      executor.submit(() -> {
        System.out.println("Executing task " + taskId + " Thread: " +
        Thread.currentThread().getName());
        try {
          TimeUnit.SECONDS.sleep(1);
        } catch (InterruptedException e) {
          Thread.currentThread().interrupt();
          System.out.println("Task " + taskId + " was interrupted");
        }
      });
```

```
    }
  executor.shutdown();

  try {
    if (!executor.awaitTermination(60, TimeUnit.SECONDS)) {
      executor.shutdownNow();
      System.out.println("Executor did not terminate in the
      specified time.");
      if (!executor.awaitTermination(60, TimeUnit.SECONDS))
        System.err.println("Pool did not terminate");
    }
  } catch (InterruptedException ie) {
      executor.shutdownNow();
      Thread.currentThread().interrupt();
  }
  System.out.println("Finished all threads");
  }
}
```

This sample application provides an efficient method of managing concurrency in a Java application.

## Java concurrency utilities

We are fortunate that the java.util.concurrent package contains a set of utilities we can use for **concurrency**. These utilities empower us to write thread-safe applications that are reliable and scalable. The utilities help us address common pitfalls and challenges in concurrent programming, including data consistency, life cycle management, and synchronization.

The advantages of using the concurrency utilities that are part of the java.util.concurrent package includes making our applications more reliable, making them perform at a higher level, and simplifying the programming effort. Let's look at five specific concurrency utilities.

Our first concurrency utility is the Executor framework that we previously covered. Looking at this framework closely reveals that there are multiple types of Executor services, including the ScheduledExecutorService interface. This interface can be used to introduce an execution delay. The primary interface, ExecutorService, empowers us to manage thread termination and helps us track our synchronous tasks.

**Synchronizers** are another set of utilities that help us synchronize threads. These methods include CountDownLatch, CyclicBarrier, Exchanger, Phaser, and Semaphore. If you need to improve your thread management in your Java application, these methods are worth reviewing in the Java documentation.

Three additional concurrency utilities are **atomic variables, locks,** and **concurrent collections**. Atomic variables provide an efficient method to perform a small operation on a single variable without the need for synchronization. This is available in `java.util.concurrent.atomic package`. Locks, available in the `java.util.concurrent.locks` package, allow us to lock threads and waits until a specific condition is met. Lastly, concurrent collections provide thread-safe collections that have full support for concurrency.

Now that we explored caching strategies, thread management, and the effective use of Java concurrency utility, you should be well-equipped to continue building high-performance Java applications.

## Summary

This chapter took a deep look at the complexities of managing memory effectively to help prevent memory leaks. These leaks must be avoided at all costs because they can degrade our systems, ruin the user experience, and even result in system crashes. We identified that memory leaks typically occur due to improper referencing, which inhibits the garbage collector's ability to deallocate memory. We focused on proper referencing, listeners and loaders, and caching and threads. You should now be equipped and confident to implement efficient memory leak avoidance strategies in your Java applications.

In the next chapter, *Concurrency Strategies*, we will cover the concepts of threads, synchronization, volatile, atomic classes, locks, and so on. We will leverage the thread-related content covered in this chapter to give us a head start as we dive deeper into concurrency. Through a hands-on approach, you can gain insights into concurrency in Java and adopt strategies to help make your Java programs highly performant.

# Part 3:
# Concurrency and Networking

Concurrency and networking are essential for modern Java applications, especially those requiring high throughput and low latency. This part introduces advanced concurrency strategies to manage multiple threads efficiently. It also covers connection pooling techniques to optimize network performance and explores the intricacies of hypertext transfer protocols. By understanding and applying these concepts, you will create highly responsive and scalable applications.

This part has the following chapters:

- *Chapter 9, Concurrency Strategies and Models*
- *Chapter 10, Connection Pooling*
- *Chapter 11, Hypertext Transfer Protocols*

# 9

# Concurrency Strategies and Models

Today's computing world consists of distributed systems, cloud-based architectures, and hardware accelerated by multi-core processors. These characteristics necessitate concurrency strategies. This chapter provides foundational information on concurrency concepts and provides hands-on opportunities to implement concurrency in Java programs. The underlying goal is to harness the benefits and advantages of concurrency to improve the performance of our Java applications.

The chapter starts with a review of different concurrency models and their practical uses. Concepts include the thread-based memory model and the message-passing model. Then, we will explore multithreading from both a theoretical perspective and a hands-on practical aspect. The thread life cycle, thread pools, and other related topics will be covered. Synchronization will also be covered to include how to ensure thread safety and strategies to avoid common pitfalls. Finally, the chapter introduces non-blocking algorithms, an advanced concurrency strategy, to improve application performance through atomic variables and specific data structures.

In this chapter, we're going to cover the following main topics:

- Concurrency models
- Multithreading
- Synchronization
- Non-blocking algorithms

By the end of this chapter, you should have a thorough understanding of concurrency and related strategies and models. You should be prepared to implement concurrency in your Java applications, ensuring they perform at a high level.

# Technical requirements

To follow the examples and instructions in this chapter, you will need the ability to load, edit, and run Java code. If you have not set up your development environment, refer to *Chapter 1*.

The finished code for this chapter can be found here: `https://github.com/PacktPublishing/High-Performance-with-Java/tree/main/Chapter09`.

# Concurrency models

One of the most exciting aspects of the Java programming language is its robustness. When addressing parallel execution challenges, Java supports multiple models, so the approach we take is up to us. Usually, there is not just one way of doing things, with each possible solution presenting both advantages and trade-offs. Our goal is to create Java applications that run efficiently and are scalable and maintainable. To that end, we will use the thread-based concurrency approach (detailed later in this chapter). Our selection is based on its straightforward nature.

**Concurrency**, in the context of computer science, is the simultaneous execution of instructions. This is achieved through multithreading (think of multitasking). This programming paradigm includes the ability to access Java objects and other resources from multiple threads. Let's look at three specific models (thread-based, message passing, and reactive) and then compare them to see which model might be more ideal, given a specific scenario, than others.

## Thread-based model

The **thread-based model**, also referred to as the **shared memory model**, is the most used Java concurrency model. The notion of a **thread** is that all threads share the same physical memory and conduct their intercommunication via that memory. The Java language has deep support for threads, with the `Thread` class and `Callable` and `Runnable` interfaces.

Let's look at a simple implementation example. We will implement the `increment` method and mark it with the `synchronized` keyword. This tells Java to only execute one thread at any given time:

```
public class MyCounter {
  private int count = 0;

  public synchronized void increment() {
    count++;
  }

  public int getCount() {
    return count;
  }
}
```

This next section of code contains our `main()` method. In this method, we create two threads; both will increment our counter:

```java
public static void main(String[] args) throws InterruptedException {
    MyCounter counter = new MyCounter();

    Thread t1 = new Thread(() -> {
        for(int i = 0; i < 1000; i++) {
            counter.increment();
        }
    });
    Thread t2 = new Thread(() -> {
        for(int i = 0; i < 1000; i++) {
            counter.increment();
        }
    });
```

The next two lines of code start the threads:

```java
    t1.start();
    t2.start();
```

In the next two lines of code, we wait for both threads to finish:

```java
    t1.join();
    t2.join();
```

Lastly, we output the final results:

```java
    System.out.println("Final counter value: " + counter.getCount());
    }
}
```

The straightforward nature of thread-based model implementation represents a tremendous advantage. This approach is typical for smaller applications. There are potential disadvantages to using this model, as **deadlocks** and **race conditions** can be introduced when multiple threads attempt to access shared, mutable data.

> **Deadlocks and race conditions**
>
> Deadlocks occur when two threads wait for the other to release a needed resource. Race conditions occur when the sequence of the thread execution is required.

Both deadlocks and race conditions should be avoided as much as possible in our applications.

# The message passing model

The **message passing model** is an interesting one in that it avoids **shared states**. This model requires threads to intercommunicate by sending messages.

> **Shared states**
>
> A shared state exists when more than one thread in an application can simultaneously access data.

The message passing model offers assurances against deadlocks and race conditions. A benefit of this model is that it promotes scalability.

Let's look at how we can implement the message passing model. Our example includes a simple sender and receiver scenario. We start with our `import` statements and then create a `Message` class:

```java
import java.util.concurrent.ArrayBlockingQueue;
import java.util.concurrent.BlockingQueue;

public class MessagePassingExample {
  static class Message {
    private final String content;

    public Message(String content) {
      this.content = content;
    }

    public String getContent() {
      return content;
    }
  }
}
```

Next, we will have our `Sender` class implement the `Runnable` interface:

```java
static class Sender implements Runnable {
  private final BlockingQueue<Message> queue;

  public Sender(BlockingQueue<Message> q) {
    this.queue = q;
  }

  @Override
  public void run() {
    // Sending messages
    String[] messages = {"First message", "Second message", "Third
    message", "Done"};
```

```
    for (String m : messages) {
      try {
        Thread.sleep(1000); // Simulating work
        queue.put(new Message(m));
        System.out.println("Sent: " + m);
      } catch (InterruptedException e) {
          Thread.currentThread().interrupt();
      }
    }
  }
}
```

Next, we will have our `Receiver` class implement the `Runnable` interface:

```
static class Receiver implements Runnable {
  private final BlockingQueue<Message> queue;

  public Receiver(BlockingQueue<Message> q) {
    this.queue = q;
  }

  @Override
  public void run() {
    try {
      Message msg;
      // Receiving messages
      while (!((msg = queue.take()).getContent().equals("Done"))) {
        System.out.println("Received: " + msg.getContent());
        Thread.sleep(400); // Simulating work
      }
    } catch (InterruptedException e) {
        Thread.currentThread().interrupt();
    }
  }
}
```

The last step is to create our `main()` method:

```
public static void main(String[] args) {
  BlockingQueue<Message> queue = new ArrayBlockingQueue<>(10);
  Thread senderThread = new Thread(new Sender(queue));
  Thread receiverThread = new Thread(new Receiver(queue));

  senderThread.start();
```

```
        receiverThread.start();
    }
}
```

Our example implemented `Sender` and `Receiver` as `Runnable` classes. They communicated using `BlockingQueue`. The queue is used for `Sender` to add messages and `Receiver` to take and process them. `Sender` sends `Done` to the queue so that `Receiver` knows when it can stop processing. The message passing model is often used in distributed systems, due to its support of highly scalable systems.

## The Reactive model

The **Reactive model** is newer than the last two models we covered. Its focus is on **non-blocking, event-driven programming**. This model is usually evident in large-scale systems that process extensive input/output operations, especially when high scalability is needed. There are external libraries that we can use to implement this model, including **Project Reactor** and **RxJava**.

Let's look at a simple implementation example using **Project Reactor**. We start by adding the Project Reactor **dependency** to our project. Here is how that looks using **Maven** as the build tool:

```
<dependency>
    <groupId>io.projectreactor</groupId>
    <artifactId>reactor-core</artifactId>
    <version>3.4.0</version>
</dependency>
```

The following example demonstrates how to create a reactive stream to process a sequence of events:

```
import reactor.core.publisher.Flux;

public class ReactiveExample {

    public static void main(String[] args) {
        Flux<String> messageFlux = Flux.just("Hello", "Reactive",
        "World", "with", "Java")
                .map(String::toUpperCase)
                .filter(s -> s.length() > 4);

        messageFlux.subscribe(System.out::println);
    }
}
```

The Reactive model offers efficient resource use, blocking operation avoidance, and a unique approach to asynchronous programming. However, it can be more difficult to implement compared to the other models we covered.

---

**Comparative analysis**

Each of the three concurrency models offers different benefits, and understanding their individuality and differences can help you make an informed decision regarding which model to adopt.

---

# Multithreading

**Multithreading** is simply the **synchronous execution**, or **concurrent execution,** of two or more parts of a program, and it is a fundamental aspect of Java's concurrent programming mechanism. We execute multiple parts of our programs, taking advantage of multi-core **Central Processing Unit (CPU)** resources to optimize the performance of our applications.

Before we get too far into multithreads, let's focus on a single **thread**. In Java, a thread is a finite unit of execution within a method, function, or process. The **Java Virtual Machine (JVM)** handles the management of our threads. As you can see in the following code snippet, we extend the `Thread` class to create and start threads:

```
class MyThread extends Thread {
    public void run() {
        System.out.println("Thread executed by extending Thread
        class.");
    }
}


// This is how we create and start a thread.
MyThread thread = new MyThread();
thread.start();
```

The next code snippet demonstrates how to implement the `Runnable` interface:

```
class MyRunnable implements Runnable {
    public void run() {
        System.out.println("Thread executed by implementing Runnable
        interface.");
    }
}
// Here, we create and start a thread.
Thread thread = new Thread(new MyRunnable());
thread.start();
```

Now that you understand how easy it is to create and start threads, let's examine their life cycles.

# Thread life cycles

Java threads have a definitive start and end state. They have additional states, as indicated in the following diagram.

Figure 9.1 – A Java thread life cycle

It is important to understand each of these states so that we can effectively manage our threads. Let's briefly look at each state within the Java thread life cycle:

1. **New**: Threads have this state when we create them but have not started them.

2. **Runnable**: This state exists when a thread is being executed. Threads that have started and are waiting for CPU time also have this state.

3. **Blocked**: A thread is blocked from accessing a resource.

4. **Waiting/Timed Waiting**: Threads can wait on other threads. Sometimes, there can be a specific waiting time, while at other times, the wait might be indefinite.

5. **Terminated**: A thread has this state after execution completes.

It is important to understand these states, especially with applications that rely on thread communication and synchronization.

## Multithreading best practices

When working with multithreading, there are a few things we should keep in mind to ensure that our application performs as expected and that our threads are safe.

First, we want to ensure that each resource is only accessed by one thread at a time, using **synchronization**. Our goal is to prevent **race conditions**. Fortunately, Java makes this easy for us, as it has the java. util.concurrent package, which includes concurrent data structures and methods we can use for synchronization. Utilizing this package can help us implement thread safety.

Java's Object class includes the wait(), notify(), and notifyAll() methods, which can be used to empower Java threads to communicate with each other. The following example application demonstrates how those methods can be used. Our example contains a Producer thread that creates a value for consumption by a Consumer thread. We do not want both operations to take place at the same time; in fact, we want Consumer to wait for Producer to create the value. Further, Producer must wait for Consumer to receive the last value before creating a new one. The first section defines our WaitNotifyExample class:

```
Public class WaitNotifyExample {
  private static class SharedResource {
```

```
    private String message;
    private boolean empty = true;

    public synchronized String take() {
      // Wait until the message is available.
      while (empty) {
        try {
          wait();
        } catch (InterruptedException e) {}
      }
      // Toggle status to true.
      empty = true;
      // Notify producer that status has changed.
      notifyAll();
      return message;
    }

    public synchronized void put(String message) {
      // Wait until the message has been retrieved.
      while (!empty) {
        try {
          wait();
        } catch (InterruptedException e) {}
      }
      // Toggle the status to false.
      empty = false;
      // Store the message.
      this.message = message;
      // Notify that consumer that the status has changed.
      notifyAll();
    }
  }
  private static class Producer implements Runnable {
    private SharedResource resource;

    public Producer(SharedResource resource) {
      this.resource = resource;
    }

    public void run() {
      String[] messages = {"Hello", "World", "Java", "Concurrency"};
      for (String message : messages) {
        resource.put(message);
```

```
        System.out.println("Produced: " + message);
        try {
          Thread.sleep(1000); // Simulate time passing
        } catch (InterruptedException e) {}
      }
      resource.put("DONE");
    }
  }
```

Next, we need to create our `Consumer` class and implement the `Runnable` interface:

```
private static class Consumer implements Runnable {
  private SharedResource resource;

  public Consumer(SharedResource resource) {
    this.resource = resource;
  }

  public void run() {
    for (String message = resource.take(); !message.equals("DONE");
    message = resource.take()) {
      System.out.println("Consumed: " + message);
      try {
        Thread.sleep(1000); // Simulate time passing
      } catch (InterruptedException e) {}
    }
  }
}
```

The last part of our application is the `main()` class:

```
public static void main(String[] args) {
  SharedResource resource = new SharedResource();
  Thread producerThread = new Thread(new Producer(resource));
  Thread consumerThread = new Thread(new Consumer(resource));

  producerThread.start();
  consumerThread.start();
}
}
```

Our application's output is provided here:

```
Produced: Hello
Consumed: Hello
```

```
Produced: World
Consumed: World
Produced: Java
Consumed: Java
Produced: Concurrency
Consumed: Concurrency
```

When we adhere to the best practices provided in this section, we increase the chances of having efficient multithreading, contributing to a high-performing Java application.

# Synchronization

**Synchronization** is another critical Java concept that we should grasp as we seek to fully understand concurrency. As we indicated earlier, we employ synchronization to avoid **race conditions**.

> **Race conditions**
>
> The condition when multiple threads attempt to modify a shared resource at the same time. The results of this situation are unpredictable and should be avoided.

Let's look at how we can implement synchronization in our Java applications by looking at several code snippets. First, we demonstrated adding the `synchronized` keyword to a method's declaration. This is how we can ensure that only one thread at a time can execute the method on a specific object:

```java
public class Counter {
    private int count = 0;
    public synchronized void increment() {
        count++;
    }
    public synchronized int getCount() {
        return count;
    }
}
```

We can also implement **synchronized blocks**, which are a subset of a method. This level of granularity allows us to synchronize a block without having to lock out the entire method:

```java
public void increment() {
    synchronized(this) {
        count++;
    }
}
```

Java also includes a `Lock` interface that can be used for a more refined approach to locking resources. Here's how we can implement it:

```java
import java.util.concurrent.locks.ReentrantLock;

public class Counter {
    private final ReentrantLock lock = new ReentrantLock();
    private int count = 0;

    public void increment() {
        lock.lock();
        try {
            count++;
        } finally {
            lock.unlock();
        }
    }
}
```

Java also includes the `volatile` keyword, which we can use to tell Java that a specific variable is subject to modification by multiple threads. When we declare our variables with this keyword, Java places the variable's value in a memory location accessible by all threads:

```java
public class Flag {
    private volatile boolean flag = true;

    public boolean isFlag() {
        return flag;
    }

    public void setFlag(boolean flag) {
        this.flag = flag;
    }
}
```

As you undoubtedly will come to understand, synchronization is key for successful concurrency programming in Java.

## Non-blocking algorithms

As a final concept of concurrent programming, let's look at **non-blocking algorithms**. These algorithms help us achieve thread safety without having to use the locking mechanisms we previously covered, such as `synchronized` methods and synchronized blocks. There are three types of non-blocking algorithms – **lock-free**, **wait-free**, and **obstruction-free**. Although their names are self-describing, let's take a closer look.

Modern CPUs support atomic operations, and Java includes several atomic classes that we can use when implementing non-blocking algorithms.

> **Atomic operations**
>
> These are operations that are executed by modern CPUs as a single, finite step that ensures consistency without the need for locks.

Here is a code snippet that illustrates how to use `AtomicInteger`:

```
import java.util.concurrent.atomic.AtomicInteger;

public class Counter {
    private AtomicInteger count = new AtomicInteger(0);

    public void increment() {
        count.incrementAndGet();
    }

    public int getCount() {
        return count.get();
    }
}
```

The following example demonstrates how to implement a non-blocking stack. As you will see, our stack uses atomic references, which ensures thread safety:

```
import java.util.concurrent.atomic.AtomicReference;

class Node<T> {
    final T item;
    Node<T> next;

    Node(T item) {
        this.item = item;
    }
}

public class ConcurrentStack<T> {
    AtomicReference<Node<T>> top = new AtomicReference<>();
```

```
public void push(T item) {
    Node<T> newHead = new Node<>(item);
    Node<T> oldHead;
    do {
        oldHead = top.get();
        newHead.next = oldHead;
    } while (!top.compareAndSet(oldHead, newHead));
}

public T pop() {
    Node<T> oldHead;
    Node<T> newHead;
    do {
        oldHead = top.get();
        if (oldHead == null) {
            return null;
        }
        newHead = oldHead.next;
    } while (!top.compareAndSet(oldHead, newHead));
    return oldHead.item;
}
}
```

We can gain performance advantages when we use non-blocking algorithms, especially when our application deals with high concurrency. The advantages are counterbalanced by code complexity, which can result in errors and code that is more difficult to maintain.

## Summary

This chapter focused on concurrency strategies and models, with the goal of providing insights into the concept of concurrency, the different models and strategies, and some implementation examples. We explored theoretical concepts and practical examples. The concepts covered included concurrency models, synchronization, and non-blocking algorithms. You should now have sufficient knowledge to start experimenting with code.

In the next chapter, we will explore connection pooling with a specific look at concepts, implementation, and best practices. You will have the opportunity to learn how to create and maintain a cache of database connection objects to help increase the performance of your Java applications.

# 10

# Connection Pooling

**Connection pooling** is a technique used in software development to manage database connections. These connections can be reused during program execution and conventional wisdom dictates that anything that can be reused should be created once and reused as needed. That has been the premise throughout this book as we strive to achieve higher-performing Java applications. This chapter covers the concept of connection pooling, providing fundamental principles, implementation approaches, and examples. Our coverage includes establishing connections, managing them, and terminating them when no longer needed. Best practices related to connection pooling will also be covered.

The following main topics are covered in this chapter:

- Connection pooling concepts
- Implementing connection pools
- Best practices with connection pools

By the end of the chapter, you should have a thorough understanding of connection pooling, be able to implement a connection pool, and strategically design an approach to leverage connection pools for performance enhancement.

## Technical requirements

To follow the examples and instructions in this chapter, you will need the ability to load, edit, and run Java code. If you have not set up your development environment, refer to *Chapter 1*.

The finished code for this chapter can be found here: `https://github.com/PacktPublishing/High-Performance-with-Java/tree/main/Chapter10`.

# Connection pooling concepts

Most modern systems include multiple databases, especially when **microservices architectures** are implemented. This makes the concept of connection pooling a critical component of efficient Java applications.

> **Microservices architecture**
>
> Microservices are independent components of a software system tied to a specific business function. They often have their own database so they can be decoupled from the main application and updated independently of other microservices.

The key issue is that software applications need to connect to databases and those connections draw on system resources. The concept of connection pooling is to establish a connection to the required databases and when they are no longer in use, return them to a pool. Obtaining a connection from a pool is quicker and less resource intensive than creating new connections every time a database operation is required.

The following illustration shows the process of connection pooling, which is also considered the **connection pool life cycle**.

Figure 10.1 – Connection pool life cycle

Connection pools are initialized when the application initially loads. Part of the initialization involves establishing the number of connections. We will walk through an example later in this chapter. For now, you should understand that we will use a **connection pool library** for our implementation.

The second component is **connection borrowing**. Whenever a database operation is necessary, a connection is obtained from the pool. The term "borrowing" suggests that once a connection is no longer required by a service, it is returned to the pool. That is the connection return component's segment of the life cycle. Unused connections are returned to the pool so they can be used again.

## Advantages of connection pooling

There are three primary advantages of connection pooling. First, the use of connection pooling can result in higher performance. This improvement is possible because connects are reused, resulting in faster database operations.

Another advantage of implementing connection pooling in our applications is that we are better equipped to optimize resource utilization. The number of database connections open at any given time is reduced because the unused connections are in a pool. This results in lower application and database server overhead.

An additional advantage of connection pooling is that it supports scalability. This is possible because when our applications use connection pooling, they can handle many simultaneous database operations.

## Challenges of connection pooling

There are challenges and concerns with virtually every high-performance approach we implement in our Java applications and connection pooling is no exception. There are three primary areas of concern.

First, it is critical that we establish an optimal **size** for our connection pool. If we do not permit enough connections, our applications can become sluggish or unresponsive. Bottlenecks can also occur when our connection pools are too small. On the other hand, if our connection pool is too large, we could possibly overtask database servers, leading to overall system performance degradation.

**Connection leaks** are another area of concern. It is important that we manage the connection pool life cycle, with a focus on connection borrowing and returning operations. When we fail to manage these operations properly, connection leaks are possible.

> **Connection leaks**
> Connection leaks occur when we fail to properly return connections to the pool. This can lead to the connection pool's resources being depleted.

You should now have a foundational understanding of connection pooling and its advantages and challenges. The next section walks through implementation examples.

# Implementing connection pools

We should now understand what connection pools are and the associated advantages and challenges. Let's extend our knowledge by implementing connection pools using Java. In this section, we will look at connection pool libraries, set up a connection pool, integrate our connection pool with application logic, and explore how to monitor our connection pools.

## Connection pool libraries

Once we decide to use connection pools in an application, we need to select an appropriate connection pool library. There are several connection pool libraries available to us for Java, and which one we select is based on our application's requirements. Let's look at three popular connection pool libraries.

**Apache Commons Database Connection Pooling (DBCP)** is a mature library that is considered stable and has wide applicability. As the name suggests, this is an open-source library from Apache. While this is a proven library, it is less efficient than more modern libraries.

The **C3PO (Cloud-Based Confidentiality-Preserving Continuous Query Processing)** connection pool library is another viable option. It includes a robust set of features to include automatic connection retries when connections cannot be established. This library is a bit more versatile than the Apache Commons DBCP library.

A third connection pool library option is the **Hikari Connection Pool (HikariCP)**. This is a newer library than the previous two and is lauded for its simplicity and its performance. With the goal of improving the performance of our Java applications, HikariCP is a great choice for connection pooling and is the library featured in the remainder of this chapter.

There are six primary factors you should consider when selecting a connection pool library:

- **Compatibility**: You should check to ensure the library is compatible with your version of Java as well as any database drivers or tools you plan to use.

- **Familiarity**: If you and your development team are already familiar with a specific connection pool library, you can introduce faster development and fewer bugs if you continue to use a library you are familiar with. The downside to this is that you might be sacrificing features and runtime performance for development efficiencies.

- **Features**: The full list of features should be reviewed to ensure the library you choose can do what you expect it to.

- **Maintenance**: We should always gravitate toward libraries that are maintainable.

- **Performance**: This is a paramount factor. You want to ensure that your chosen library does not underperform when under stress (high and persistent workloads). This is something you should test before formally adopting a connection pool library.

- **Support**: Check the official website to ensure there is ample documentation. Additionally, you want to select a library that has robust community support. This can help you when you experience development challenges and when troubleshooting.

When selecting a connection pool library, overall application performance should be heavily considered. This may require trial and error with multiple libraries. Reviewing the features of each library can help you make an informed decision. The following table can help with your review.

| Feature | Apache DBCP | C3P0 | HikariCP |
| --- | --- | --- | --- |
| Performance | Good | Good | Great |
| Connection timeout | Yes | Yes | Yes |
| Statement caching | Yes | Yes | Yes |

| Feature | Apache DBCP | C3P0 | HikariCP |
|---|---|---|---|
| Idle connection test/validation | Yes | Yes | Yes |
| Connection validation | Yes | Yes | Yes |
| Pool sizing flexibility | Good | Good | Great |
| Documentation | Good | Good | Great |
| Community support | Good | Good | Great |
| Configuration ease | Moderately Complex | Moderately Complex | Simple |
| Modern framework integration | Poor | Good | Great |

Table 10.1 – Library features

As you can see in the preceding table, many of the features are rated the same across all three connection libraries. This suggests that a deeper dive might be required. The comparison is only provided as a high-level overview and provides insights into areas you might consider researching further.

## Setting up a connection pool

Now that a connection pool library has been selected, HikariCP in our case, we need to follow a few specific steps. Let's walk through an example using Maven for our build tool:

1. **Add the library to your project:**

   We need to edit our pom.xml file to add HikariCP to our dependencies. This is how we would do that:

   ```
   <dependency>
       <groupId>com.zaxxer</groupId>
       <artifactId>HikariCP</artifactId>
       <version>2.6.3</version>
   </dependency>
   ```

2. **Connection pool configuration:**

   Note that the following source code is just an example and will not run on its own. It is provided to help explain how to configure a connection pool. As you can see, we create dataSource and set several parameters to configure our connection pool:

   ```
   import com.zaxxer.hikari.HikariConfig;
   import com.zaxxer.hikari.HikariDataSource;

   public class DatabaseConfig {
     private static HikariDataSource dataSource;
   ```

```
static {
  HikariConfig config = new HikariConfig();
  config.setJdbcUrl( "jdbc:postgresql://localhost:5432/
  myDatabase");
  config.setUsername("databaseUser");
  config.setPassword("databasePassword");
  // Pool configuration
  config.setMaximumPoolSize(10);
  config.setMinimumIdle(5);
  config.setIdleTimeout(600000);
  config.setMaxLifetime(1800000);
  config.setConnectionTimeout(30000);

  dataSource = new HikariDataSource(config);
  }
  public static HikariDataSource getDataSource() {
    return dataSource;
  }
}
```

3. **Pool initialization**:

In the previous code snippet, we initialized a `static` connection pool in the `DatabaseConfig` class. We implement this approach so our connection pool is initialized and ready when the class loads. The global access point to our connection is established with the `getDataSource()` method. This facilitates connections to be borrowed from the pool by our application's methods.

4. **Using the pool**:

With our connection pool configured and loaded, we are ready to start using it. Since we already implemented the `getDataSource()` method, we can access the pool and obtain connections. Here is a method of accomplishing that task. As previously stated, this is not a complete application; rather, it is a representative code snippet to demonstrate how to use a connection pool:

```
import java.sql.Connection;
import java.sql.ResultSet;
import java.sql.Statement;

public class DatabaseOperations {
  public void executeQuery(String query) {
    try (Connection conn = DatabaseConfig.getDataSource().
    getConnection();
        Statement stmt = conn.createStatement();
        ResultSet rs = stmt.executeQuery(query)) {
      while (rs.next()) {
```

```
            // Here you would process the result set
        }
    } catch (Exception e) {
        e.printStackTrace();
    }
  }
}
```

As shown previously, we implemented the `try-with-resources` statement to automatically close our database connection, which returns the connection to the pool so it can be used again.

## Integrating connection pools

Integrating connection pools into our Java applications requires us to create code (as demonstrated in the previous section) that creates, uses, and closes database connections. We highlighted that we no longer need to open a new database connection every time a database connection is required. Instead, we borrow from the pool and return our connections when we are done with them.

The primary integration points are as follows:

- Managing connections
- Obtaining connections
- Handling errors

Once we have our connection pool in place, we need to monitor them during runtime and perform tuning as needed. Let's look at those tasks in the next section.

## Monitoring connection pools

Connection pools are an important component of modern software systems, and they can represent a tremendous boon to overall system performance. This underscores the importance of monitoring their performance during runtime. We can accomplish this by reviewing logs and using monitoring tools. Most connection pool libraries come with tools sufficient for the task.

In addition to errors, we should look for the following:

- Connection leaks
- Long wait times
- Improper pool sizes

Part of monitoring connection pools includes the need to establish key metrics. Simply watching system logs is not enough; we need a set of metrics or benchmarks to properly measure the success and performance of our connection pools.

As we continue to monitor our connection pool performance, we can make configuration changes that are applied when the application starts up. For always-on systems, you may need to restart a service or server. A mindset of continual performance improvement can help ensure we get the best possible performance from our connection pools and positively impact our system's overall performance.

# Best practices with connection pools

Connection pool implementation is relatively straightforward and, like most programming tasks, you will quickly have your own code base that you can refactor for follow-up programming projects. This is often a critical component to your system's success as it provides the mechanism for your application to access data. Some factors should be considered as part of your connection pool strategy. Let's look at the primary factors.

## Connection pool sizing

Ensuring your connection pool is properly sized is the first factor you should consider. We should strive to find the ideal balance between performance and resource usage. If our pool is too small, the access wait times can increase, which will negatively impact performance. Oversized connection pools can result in wasted resources such as idle connections requiring system resources.

The challenge is knowing what the ideal connection pool size is. This can take some trial and error. Initially, we can estimate the number of database connections our application might need at one time. There is no magic formula for this, so consider the following when making your initial estimates:

- The number of services that your application has that need a database connection to fully function
- The number of concurrent connections you will need
- Review the usage patterns
- The peak load conditions

Using our `HikariCP` example from earlier, we can set the size of our connection pool with a single line of code in our `DataSourceConfig` class:

```
config.setMaximumPoolSize(10);
config.setMinimumIdle(5);
```

As you can see, we set the maximum number of connections to 10 and the minimum number of idle connections that the pool maintains to 5.

Once you make your initial connection pool size decision, continue monitoring the performance and adjusting your configuration as needed. You can use tools that come with your connection pool library as well as external tools such as **application performance monitoring** (APM) tools.

# Handling connection leaks

Once your application is running, you should commit yourself to continually monitoring your connection pool performance. While you hope not to experience connection leaks, the best practice is to be prepared for them. As a reminder, a connection leak occurs when a connection is not returned to the pool after it is no longer in use. This can lead to depleting the available connections in the pool. Ultimately, this can lead to your application failing.

There are two primary approaches to handling connection leaks, and they can be used in concert with one another.

## Timeout settings

We can set a timeout period for connections that are borrowed. If a connection has been borrowed from the pool for what you consider to be too long, then we can have it reclaimed or at least a log entry made to help with your monitoring efforts.

Let's review the pool configuration part of the code snippet from earlier in this chapter:

```
config.setIdleTimeout(600000);
config.setMaxLifetime(1800000);
config.setConnectionTimeout(30000);
```

As seen in the preceding code snippet, we set the maximum amount of time a connection can be in the pool, in an idle state to 60,000 milliseconds (about 1 minute). The second line of code sets the maximum lifetime of a connection in the pool to approximately 30 minutes, and the third line sets the amount of time to wait for a connection from the pool.

## Connection handling patterns

We should review our code to ensure connections are always closed. This can be done in a `finally` block or in the `try-with-resource` statement we used earlier in this chapter. Let's look at an example of each.

The first code snippet is in an abbreviated format for illustration purposes only. It demonstrates how we can ensure the connections are closed. In the following example, we assume that all appropriate import statements will be included and that `dataSource` has been initialized elsewhere in the application. In the `finally` block, we ensure that resources are closed to help avoid connection leaks:

```
// import statements
public class DatabaseUtil {
    private DataSource dataSource;
    public void executeQuery(String query) {
        Connection conn = null;
        Statement stmt = null;
```

```
        ResultSet rs = null;
        try {
            conn = dataSource.getConnection();
            stmt = conn.createStatement();
            rs = stmt.executeQuery(query);
            // Process the result set
            while (rs.next()) {
                // Handle data
            }
        } catch (Exception e) {
            // Handle exception
        } finally {
            if (rs != null) {
                try {
                    rs.close();
                } catch (Exception e) {
                    e.printStackTrace();
                }
            }
            if (stmt != null) {
                try {
                    stmt.close();
                } catch (Exception e) {
                    e.printStackTrace();
                }
            }
            if (conn != null) {
                try {
                    conn.close();
                } catch (Exception e) {
                    e.printStackTrace();
                }
            }
        }
    }
}
```

While implementing the `finally` block can help ensure that resources are closed to help avoid connection leaks, another approach is to use the `try-with-resources` statement. Here is an example of that statement:

```
public void executeQuery(String query) {
    try (Connection conn = dataSource.getConnection();
        Statement stmt = conn.createStatement();
```

```
                ResultSet rs = stmt.executeQuery(query)) {
                // Process the result set
                while (rs.next()) {
                    // Handle data
                }
        } catch (Exception e) {
            // Handle exception
            e.printStackTrace();
        }
    }
}
```

This approach results in the resources automatically being closed once the try-catch block is exited. This exit can occur based on normal program flow or when an exception is caught. In both cases, the resource will be automatically closed. As you saw, there was no need for a `finally` block with this approach and it is therefore the recommended approach.

## Connection pool security

Connection pools facilitate access to our databases, and we must always protect them. Maintaining a pool of database connections could represent a tremendous security risk. There are two types of protection we can implement.

First, we can encrypt our configuration files. These files contain our database connection information and should be considered sensitive information. Encryption and decryption can take processing time resulting in a small downtick in performance, but this is a necessary component of your application.

Another type of protection is to use the concept of **least privilege**, only granting the minimum privileges necessary for the application to run. For example, if you have a service that merely needs to search a customer database to display key information such as names, emails, and account numbers, do not give that service create, update, or delete access to the database. In this scenario, you only need to grant the service read access.

## Advanced topics

There are a few connection pool topics that go beyond the basics and are worth considering as we aim to optimize the performance of our connection pools. Let's look at four of those topics here:

- **Cloud native:** When we are working with cloud-based applications, we can leverage features native to the cloud environment. This can include features aimed at improving resiliency, reliability, and scalability. Ideally, our database selection will be based on cloud-native database services to further optimize connection pooling.

- **Connection validation:** It is a good idea to periodically execute a function to validate the connections in the pool. This can ensure they remain valid and can prevent costly problems.

- **Failover**: Database redundancy is a key feature of cloud computing environments. Specific to connection pools, we can implement a failover schema that shifts to a backup database if the first one fails.

- **Redundancy**: To support failover, and as part of normal practice, we should implement database redundancy. Taking advantage of cloud computing database services can make this relatively easy to configure.

Following the best practices presented in this section can help you implement connection pools in a manner that contributes to the high performance of your Java applications.

## Summary

This chapter took a deep look at the foundational concepts and components of connection pools with a focus on increasing the performance, resilience, reliability, and scalability of our Java applications. We also explored implementation strategies and best practices for optimizing our use of connection pools. Specifically, we highlighted how connection pools work, their advantages, and related challenges for developers. We reviewed the features of multiple connection pool libraries and selected the HikariCP library for our code examples. You should have a firm understanding of connection pools, why we should use them, and how to create, monitor, and fine-tune them.

In the next chapter, we will look at the **Hypertext Transfer Protocol** (**HTTP**). This protocol is used to transmit data and is the backbone of data communication over the web. Our focus will be on how to leverage HTTP for our Java applications to communicate with web browsers and web servers. The chapter aims to help you learn how to use HTTP in Java web applications while maintaining high performance, how to implement strategies for effectively using HTTP with Java, and how to use HTTP to communicate between Java applications and APIs.

# 11

# Hypertext Transfer Protocols

**Hypertext Transfer Protocol (HTTP)** is the foundational protocol used for information exchange on the web. It empowers communication between client computers and servers. HTTP's applicability to Java is primarily due to Java web applications.

This chapter starts with an introduction to HTTP within the context of Java. Once the fundamentals have been addressed, the chapter moves on to a practical approach to using built-in and third-party HTTP-related libraries. Using HTTP for API integrations is also covered. The chapter also looks at security concerns with using HTTP and touches on the use of HTTPS. The concluding section focuses on performance optimizations for our Java applications when employing HTTP. By the end of this chapter, you should have a firm understanding of how HTTP impacts performance in Java applications and be comfortable with your future HTTP implementations.

This chapter covers the following main topics:

- Introduction to HTTP
- Java web applications
- Using HTTP in Java
- API integration
- Security considerations
- Performance optimization

## Technical requirements

To follow the examples and instructions in this chapter, you will need to have the ability to load, edit, and run Java code. If you have not set up your development environment, refer to *Chapter 1, Peeking Inside the Java Virtual Machine (JVM)*.

The finished code for this chapter can be found here: `https://github.com/PacktPublishing/High-Performance-with-Java/tree/main/Chapter11`.

# Introduction to HTTP

HTTP was released around 1990, so you are likely familiar with at least the basic concept of enabling client and server communication. The World Wide Web was built on HTTP, and HTTP is still the foundation for the web today.. Understanding this protocol is important if you plan to build a web application, especially in Java. We will explore HTTP in the context of Java in this section.

## HTTP core components

HTTP is a stateless protocol that uses a communication model called **request-response**. Each request-response exchange is independent of all others, making this a very simple protocol. Today's web services rely heavily on HTTP.

> **Stateless protocol**
>
> A stateless protocol is one where the receiver does not need to retain or track information about the information being sent. No session information is saved.

There are four basic components of HTTP:

- First is the **request and response pair**. This pair represents the core of all web communication. A client sends a request for information, such as loading a web page, and the server sends back either a response with the requested information or another action result such as an error.

- Both the requests and responses contain **headers** that provide metadata. This HTTP component includes information such as the type of content being sent, what was requested, authentication details, and other information.

- A third HTTP component is **status codes**. When servers respond to a client's request, they provide a status code to characterize the outcome of the request. These codes are categorized by a numbered series as indicated in the table that follows:

| Category | Type of response | Example |
|---|---|---|
| 100 series | Informational responses | 100 Continue |
| 200 series | Successful responses | 200 OK |
| 300 series | Redirection messages | 301 Moved permanently |
| 400 series | Client errors | 404 Not found |
| 500 series | Server errors | 500 Internal server error |

Table 11.1 – HTTP status codes and examples

- The fourth HTTP component is **methods**. There are several request methods that HTTP uses to perform required actions. Here is a list of the most common HTTP methods:

| Method | Functionality |
|--------|---------------|
| DELETE | Deletes a resource |
| GET | Retrieves a resource |
| HEAD | Retrieves a resource's header |
| POST | Submits data to a resource |
| PUT | Updates a resource |

Table 11.2 – HTTP methods

Now that we have a fundamental understanding of HTTP, let's examine the significance of this protocol to Java developers.

## Java and HTTP

There are a plethora of libraries and **Application Programming Interfaces** (**APIs**) to help us manage how our apps use HTTP. HttpClient, for example, helps simplify our use of HTTP operations. Learning how to use available APIs and libraries is important, especially when we are concerned with how our Java applications perform.

One of the reasons HTTP knowledge is so important is that most API integrations involve HTTP communications. This requires us to understand how to formulate an HTTP request, how to handle response status codes, and how to parse the responses.

Java developers should also master HTTP when developing web applications. HTTP is the underlying protocol used for client-server communications. This underscores the importance of HTTP knowledge for Java developers.

Another reason why Java developers should seek to fully understand HTTP is that it plays a significant role in overall program performance. HTTP lacks complexity but is nevertheless a critical protocol for developing web apps, microservices, applets, and other application types.

# Java web applications

A Java web application is a server-side application used to create dynamic websites. We create websites that interact with Java web applications to dynamically generate content based on user input. Common examples include the following:

- E-commerce platforms
- Enterprise applications

- Online banking
- Information management systems
- Social media platforms
- Cloud-based applications
- Educational platforms
- Healthcare applications
- Gaming servers
- **Internet of Things (IoT)** applications

These examples demonstrate the versatility of using Java to develop and deploy high-performance web applications. As you might expect, HTTP is a foundational component of these Java web applications.

Let's next review the basic architecture for Java web applications, so we can understand HTTP's role.

## Java web application architecture

Most Java web applications are comprised of four tiers, making it a multi-tier architecture:

- **Client tier**: The client tier is what the user sees, normally via a web browser. These web pages usually consist of **Hypertext Markup Language (HTML)**, **Cascading Style Sheets (CSS)**, and **JavaScript (JS)**.
- **Web tier (or server tier)**: This tier receives and processes the HTTP requests. We can use several technologies such as **JavaServer Pages (JSP)** to accomplish this.
- **Business tier**: The business tier is where our application logic resides. This is where data is processed, computations are performed, and logic-based decisions are made. The link between this tier and the web tier is extensive.
- **Data tier**: The data tier is a critical part of the backend system. This tier is responsible for managing databases and ensuring data security and persistence.

## Key technologies

Key technologies worth mentioning are serverlets, JSPs, Spring Framework, and Jarkata EE:

- **Serverlets**: Java programs that run on web servers are referred to as serverlets. This specialty software sits between the client's HTTP requests and the applications and/or databases on the web servers.
- **JSP**: JSPs are text documents used to execute on the server and generate content for dynamic web pages. JSPs are typically used in conjunction with serverlets.

- **Spring Framework**: Spring is a Java application framework that is commonly used for developing Java web applications.

- **Jakarta EE**: Jakarta Enterprise Edition (**Jakarta EE**) is a set of application specifications that extend the Java **Standard Edition** (**SE**). It includes specifications for web services and distributed computing.

## Steps for creating a simple Java web application

There are six basic steps to creating Java web applications. Please note that this is an abbreviated approach and will vary based on your application needs, such as the need for databases, APIs, and so on:

1.  The first step is to establish your development environment. This will consist of an **Integrated Development Environment** (**IDE**) such as Visual Studio Code, the most recent **Java Software Development Kit** (**JDK**), and a web server such as Apache Tomcat.

2.  The next step is to create a new web application project in your IDE. This step includes creating the project's file directory structure.

3.  Next, we will write the code for our Java serverlet to handle HTTP requests. During this step, we will also define the routes that our serverlets will respond to. This is often defined using URLs.

4.  Next, we will be creating the JSP pages or templates we plan to use to generate HTML content. This is the content we will send back to the client.

5.  Next, we will create the business logic to implement at the core of our application. We can accomplish this through a series of Java classes.

6.  Lastly, we package our application into a **Web Application Archive** (**WAR**), which is like a **Java Application Archive** (**JAR**) but for web applications, and deploy it.

When developing Java web applications, we should create them with distinct boundaries between the presentation, business, and data access layers. This approach will help with modularity, scalability, and maintainability. It is also advisable to use frameworks such as Spring and Jakarta EE. Doing so can simplify our development efforts and provide inherent support for web application development.

Dynamic web pages are the norm and expected by users, so embracing Java web application technologies is important for all Java developers.

# Using HTTP in Java

As Java developers, we can use HTTP to create dynamic web applications. These will use HTTP to communicate between the browser and server. Java includes `HttpClient`, a Java library that makes working with HTTP requests and processing responses efficient. Let's look at an example.

The preceding code employs the `HttpClient` library to create a GET request, which retrieves data from a specific resource (simulated in our example):

```java
import java.io.IOException;
import java.net.URI;
import java.net.http.HttpClient;
import java.net.http.HttpRequest;
import java.net.http.HttpResponse;

public class GetRequestExample {
  public static void main(String[] args) {
    HttpClient client = HttpClient.newHttpClient();
    HttpRequest request = HttpRequest.newBuilder()
        .uri(URI.create("https://api.not-real-just-
        an-example.com/data"))
        .GET()
        .build();
    try {
      HttpResponse<String> response = client.send(request,
      HttpResponse.BodyHandlers.ofString());
      System.out.println("Status Code: " +
      response.statusCode());
      System.out.println("Response Body: \n" + response.body());
    } catch (IOException | InterruptedException e) {
        e.printStackTrace();
    }
  }
}
```

The preceding example sends a GET request to a simulated URL and prints both the status code and the body of the response.

Next, let's look at a method for making a POST request. This type of request can be used to submit data to a specific resource using JSON. In our example, this will be a simulated resource:

```java
import java.io.IOException;
import java.net.URI;
import java.net.http.HttpClient;
import java.net.http.HttpRequest;
import java.net.http.HttpResponse;
import java.net.http.HttpHeaders;
```

```
import java.nio.charset.StandardCharsets;

public class PostRequestExample {
  public static void main(String[] args) {
    HttpClient client = HttpClient.newHttpClient();
    String json = "{\"name\":\"Neo
    Anderson\",\"email\":\"neo.anderson@thematrix.com\"}";
    HttpRequest request = HttpRequest.newBuilder()
        .uri(URI.create("https://api.not-real-just-an-
        example.com/users"))
        .header("Content-Type", "application/json")
        .POST(HttpRequest.BodyPublishers.ofString(json,
        StandardCharsets.UTF_8))
        .build();

    try {
      HttpResponse<String> response = client.send(request,
      HttpResponse.BodyHandlers.ofString());
      System.out.println("Status Code: " +
      response.statusCode());
      System.out.println("Response Body: \n" +
      response.body());
    } catch (IOException | InterruptedException e) {
        e.printStackTrace();
    }
  }
}
```

This example simply sends a POST request to a simulated URL with a JSON package that contains user information.

Making use of the HttpClient library can simplify the process of developing code that interacts with web services.

# API integration

When we build Java web applications, we can integrate with external APIs to extend the functionality of our applications. An example would be a weather service API that can be used to display the local temperature on a site. For this section, we will focus on the **Representational State Transfer (RESTful)** services because those are the most common type of web API.

The RESTful APIs use standard HTTP methods, such as the GET and POST examples from the previous section. As you would expect, RESTful APIs communicate primarily through HTTP data exchanges using JSON and XML formats.

When we implement an API, we first learn what its required request methods are, as well as the prescribed format for requests and responses. It is increasingly common for APIs to require authentication, so that might be something you will need to contend with using API keys or other authorization techniques.

The example that follows demonstrates a simple application that implements a weather API:

```java
import java.io.IOException;
import java.net.URI;
import java.net.http.HttpClient;
import java.net.http.HttpRequest;
import java.net.http.HttpResponse;

public class WeatherApiExample {
  public static void main(String[] args) {
    HttpClient client = HttpClient.newHttpClient();
    String apiKey = "your_api_key_here";
    String city = "Florence";
    String uri = String.format("https://api.fakeweatherapi.com/v1/
    current.json?key=%s&q=%s", apiKey, city);

    HttpRequest request = HttpRequest.newBuilder()
      .uri(URI.create(uri))
      .GET()
      .build();
    try {
      HttpResponse<String> response = client.send(request,
      HttpResponse.BodyHandlers.ofString());
      System.out.println("Weather Data: \n" + response.body());
    } catch (IOException | InterruptedException e) {
        e.printStackTrace();
    }
  }
}
```

Our example sends a GET request to an API, passing the city as a query parameter. The JSON response would contain the applicable weather data, which is printed on the system's display.

API integration can be considered a core component of many Java web applications based on its wide applicability.

## Security considerations

Whenever we add functionality to our Java applications that sends information external to our application or receives information from external sources, security becomes a paramount concern. This is especially true when we integrate APIs and HTTP into our Java applications. Let's look at nine best practices that we can use to help ensure our HTTP communications are secure as well as when working with APIs:

- **Use HTTPS instead of HTTP**: If your Java web application handles sensitive, protected, or private information, you should use **HTTP Secure** (**HTTPS**) instead of HTTP when transmitting requests and responses. This will help prevent tampering and data interception. This will require you to obtain **Secure Sockets Layer** (**SSL**) certificates for your servers.

- **Do not trust input**: We should always validate input to our systems to include user input and data passed to our applications programmatically. We should not assume that this data is in the right format. After we validate the data, we may have to clean it so it can be used in our application. This approach can help mitigate nefarious operations such as **SQL injections**.

- **Authenticate**: Whenever possible, identify the users and systems that your application interacts with. The previously mentioned API keys come into play here.

- **Authorize**: Once a user or system has been authenticated, we should ensure that they have permission to perform specific operations in your application. Not every user will have the same level of authority.

- **Protect API keys**: We have already mentioned the importance of API keys and their applicability to addressing security concerns. API keys are like passwords; we must protect them from exploitation. We do not want to hardcode these keys in our applications; instead, we should store them in encrypted configuration files, so they are protected from unauthorized eyes.

- **Use security headers**: We have the option of using **HTTP security headers**. Here are some details:

  - **Content Security Policy** (**CSP**): This helps prevent XSS attacks by explicitly identifying resources that are permitted to be loaded

  - **HTTP Strict Transport Security** (**HSTS**): This can be used to enforce secure server connections

- **Treat sensitive data carefully**: This should go without saying, but sensitive data deserves special attention. For example, never transmit sensitive data in URLs, because they can be logged and then leaked. Also, ensure that sensitive data (such as passwords) is encrypted or hashed when you store it. Additionally, use a technique such as **tokenization** to securely handle payment information.

- **Update dependencies**: We should periodically check that our dependencies and Java libraries are up to date. We do not want to use older versions of components that might have known vulnerabilities.

- **Log and monitor**: As with all of our software, we want to ensure we implement proper logging and then monitor operations to ensure the logs do not contain sensitive information.

Security should always be at the forefront of developers' minds. It is especially important when working with HTTP and external APIs. Adhering to the nine best practices discussed in this section is a good start to developing a security strategy for your Java web applications.

# Performance optimization

Now that we have sufficiently covered what HTTP is, its applicability to Java, and some techniques and best practices, let's consider performance-related issues specific to using HTTP with Java. Our goals for looking at performance issues are to enhance the user experience and improve our application's scalability, resilience, and maintainability.

Let's look at seven specific areas regarding performance optimization when using HTTP in our Java applications:

- The first area focuses on the **HTTP client**. When using HTTP clients, we want to use them efficiently. Here are three techniques for doing that:

  - We discussed connection pooling in *Chapter 10, Connection Pooling*, and can apply those same principles to HTTP connections. This can help reduce the computational overhead and improve performance. The `HttpClient` library includes support for connection pooling.

  - We can use `HTTP Keep-Alive` to keep connections open for multiple requests to a common host. This will reduce the number of communication handshakes and improve latency.

  - We can often leverage asynchronous requests (that is, API calls) to improve application flow.

- **Caching** is another area to look at to help optimize performance. There are a few caching strategies that can be used to improve performance:

  - Cache at the application level for frequently accessed data. The specifics depend on your application and what data it uses. There are even caching frameworks such as Caffeine that can be used.

- Using HTTP caching headers (that is, Cache-Control) can help you control response caching.

- If your Java web application deals with static content (that is, images), you can consider using **Content Delivery Networks (CDNs)** to cache your content closer to your users (i.e., storing data on servers in specific geographic areas). This approach can significantly shorten load times for users.

- A third area to consider is **optimizing data transfers**. There are two specific approaches to improving data transfers that are worth considering:

  - To the fullest possible extent, we should minimize data requests. Obviously, the fewer data requests there are, the better our applications will perform. Achieving this takes a purposeful approach to API integration design. We can use specific API endpoints to only obtain data necessary for the task instead of a bloated package.

  - There are data compression tools we can use to decrease the size of the HTTP responses. This approach has become commonplace, so your web server is apt to support this type of compression.

- **API performance** is a fourth area of concern. Here are two techniques for optimizing API performance:

  - When possible and applicable, implement rate limitations on your APIs. This can help prevent abuse and denial of service attacks. It can also help maintain service quality.

  - If your application and APIs support batch requests, it is worth implementing. This can have a profound impact on system performance.

- **Code optimization** is a fifth area of concern. Profiling tools such as VisualVM and JProfiler can be employed to help identify performance bottlenecks. The tools can be used to target memory and CPU operations. See *Chapter 14, Profiling Tools*, for more information.

- **SQL optimization** is another area of concern. SQL queries can be optimized to reduce database load and execution time. A thorough review of database schemas can help identify additional opportunities for optimization. See *Chapter 15, Optimizing Databases and SQL Queries*, for additional information.

- Our last area of performance concern when dealing with HTTP in Java is **scalability**. The two major techniques in this area are load balancing to help improve application availability and microservices architecture for better performance.

Optimizing every aspect of our Java applications is important if we are to achieve our goal of developing high-performance applications. Working with HTTP in Java represents a unique set of challenges and opportunities for optimization.

# Summary

This chapter highlighted the role that HTTP plays in Java web application development. The purpose of HTTP was stated as being to facilitate dynamic web applications. Our coverage of this topic showed that HTTP can be used efficiently and securely. We also looked at Java web applications, API integrations, security, and performance optimization strategies. Since the fields of HTTP and Java web application development are so vibrant, it is important to be aware of changes and updates as they become available.

The next chapter, *Chapter 12*, *Frameworks for Optimization*, introduces strategies for using asynchronous input/output, buffered input/output, and batch operations to create high-performance Java applications. The chapter also covers frameworks for microservices and cloud-native applications.

# Part 4:
# Frameworks, Libraries, and Profiling

Leveraging the right frameworks and libraries can greatly enhance application performance. This part examines various frameworks designed for optimization and introduces performance-focused libraries that can be integrated into Java projects. Additionally, it provides a guide to using profiling tools for the identification and resolution of performance bottlenecks. The chapters in this section are designed to equip you with the tools and knowledge needed to fine-tune your applications for maximum efficiency.

This part has the following chapters:

- *Chapter 12, Frameworks for Optimization*
- *Chapter 13, Performance-Focused Libraries*
- *Chapter 14, Profiling Tools*

# 12
# Frameworks for Optimization

Optimization frameworks are libraries, tools, and guidelines designed to help developers enhance the performance of their Java applications. Examples include streamlining processes, reducing resource utilization, and reducing processor burdens. That is the focus of this chapter. We will start by looking at asynchronous input and output, its importance, and associated libraries and frameworks. The chapter then explores buffered input and output to include use cases and performance impacts. You can learn how both asynchronous and buffered input/output operations can improve efficiency and reduce latency. The benefits of batch operations, and related frameworks and APIs, will be explored. We will review techniques to optimize batch operations, minimizing resource utilization and maximizing data flow.

The chapter introduces microservices and covers specific frameworks that can be used with microservices. These frameworks, along with those for cloud-native applications, will be explored, as these advanced architectures are pervasive in modern software development. We will highlight how those frameworks can be implemented to optimize the performance of our Java applications. To conclude the chapter, we will review several case studies and performance analyses, providing practical context to use the chapter's featured frameworks in real-world scenarios.

By the end of this chapter, you should have a foundational understanding of key frameworks to optimize Java applications. The case studies should help deepen your understanding of how these frameworks can impact application performance. You should also be comfortable creating and implementing optimization strategies, based on the frameworks presented in this chapter.

This chapter covers the following main topics:

- Asynchronous input/output
- Buffered input/output
- Batch operations
- Frameworks for microservices
- Frameworks for cloud-native applications
- Case studies and performance analysis

# Technical requirements

To follow the examples and instructions in this chapter, you will need the ability to load, edit, and run Java code. If you have not set up your development environment, refer to *Chapter 1, Peeking Inside the Java Virtual Machine (JVM)*.

The finished code for this chapter can be found here: `https://github.com/PacktPublishing/High-Performance-with-Java/tree/main/Chapter12`.

# Asynchronous input/output

**Asynchronous** refers to uncoordinated (unsynchronized) communication operations. In the context of input/output operations, data does not have to be transmitted in a steady stream. This is a technique that we can use in Java to allow our programs to handle input and output operations without blocking the main thread's execution. While it may not always be necessary to employ this technique, it can offer great performance advantages when dealing with systems that rely on high responsiveness and performance.

For brevity, let's refer to **asynchronous input/output** using the **AIO** acronym. With AIO, we can initiate processes and then have them run independently, which allows our main application to continue running other processes. Consider the **synchronous** alternative, where the main application must wait for one operation to complete before running another one. The synchronous approach can result in latency, longer response times, and other inefficiencies.

Now that you have a foundational understanding of asynchronous input/output, let's examine the advantages and best practices of using this technique to improve Java application performance.

## Advantages and best practices

There are several advantages to AIO, and they are best leveraged when following industry best practices. First, we will review the advantages, followed by the best practices.

### Advantages

Here are three advantages to implementing AIO in our Java applications:

- **Efficiency**: AIO implementation can result in resource efficiency because we can use threads for processing, instead of waiting on input/output operations.

- **Responsiveness**: When we decouple input/output operations from the main thread execution, we increase the application's overall responsiveness. With AIO, the main application can remain responsive to input (i.e., user interactions) while other input/output operations are occurring.

- **Scalability**: If your application needs to be scalable, you will most certainly want to consider implementing AIO, which is essential for building scalable applications. AIO helps us manage multiple simultaneous connections without the need for additional threads. This significantly reduces overhead.

Keeping these advantages in mind, let's review the best practices to optimize them in Java.

### Best practices

Here are several best practices to help guide your use of AIO:

- **Error handling**: Your AIO implementation strategy should include robust error handling to catch and handle exceptions.

- **Handlers**: The use of **callback handlers** to react to input/output events is advisable to help you keep your code organized and maintainable.

- **Resource management**: As with most programming, you should ensure that all resources used in your application (i.e., network sockets) are closed after their associated operation concludes. This will help prevent resource/memory leaks.

We have established that implementing AIO is an approach that can positively impact the performance of our applications. In the next section, we will look at how to implement AIO in Java.

## Implementing AIO

The Java platform includes a **New Input/Output** (**NIO**) library that includes the following capabilities:

- `AsynchronousFileChannel`: This class enables us to read and write from and to files without blocking other tasks.

- `AsynchronousServerSocketChannel` and `AsynchronousSocketChannel`: These classes are used to handle asynchronous network operations, which helps make our applications scalable.

- **Channels** and **buffers**: Java's NIO library relies heavily on channels and buffers. Channels are the connections to components that perform input/output operations, such as network sockets. Buffers handle the data.

There are several AIO-related frameworks and libraries available to us, in addition to Java NIO. Here are two examples:

- **Akka**: This is a toolkit consisting of libraries to help us build resilient Java applications, with a focus on distributed and current systems.

- **Netty**: This is a framework for high-performance applications that makes it easy for developers to create network applications. It supports both AIO and event-driven communication models.

Let's look at AIO in code. The following example application demonstrates the `AsynchronousFileChannel` class to conduct an asynchronous file read operation. As you will see, the application employs a callback mechanism to handle the read operation's completion.

Our application starts with a series of `import` statements required by it. As you can see, we are leveraging Java's NIO library:

```
import java.nio.ByteBuffer;
import java.nio.channels.AsynchronousFileChannel;
import java.nio.file.Path;
import java.nio.file.Paths;
import java.nio.file.StandardOpenOption;
import java.util.concurrent.Future;
```

Next is our `main()` class:

```
public class CH12AsyncFileReadExample {
  public static void main(String[] args) {
    Path path = Paths.get("ch12.txt");
    try (AsynchronousFileChannel fileChannel =
AsynchronousFileChannel.open(path, StandardOpenOption.READ)) {
      ByteBuffer buffer = ByteBuffer.allocate(1024);
      Future<Integer> result = fileChannel.read(buffer, 0);
      while (!result.isDone()) {
        System.out.println("Processing something else while reading
        input...");
      }
      int bytesRead = result.get();
      System.out.println(bytesRead + " bytes read");
      buffer.flip();
      byte[] data = new byte[bytesRead];
      buffer.get(data);
      System.out.println("Read data: " + new String(data));
    } catch (Exception e) {
      System.err.println("Error encountered: " + e.getMessage());
    }
  }
}
```

As we can see from the preceding code, we open the `ch12.txt` file using `AsychnronousFileChannel`. We open it asynchronously using `AsynchronousFileChannel.open()`, specify the path, and set the read-only open option. Next, we use a `ByteBuffer` to hold the data we read from the file. This is a non-blocking method, which immediately returns a `Future` object that represents the pending result. We simulate processing other tasks during the read operation and print a message from the main thread. Lastly, we implemented a while loop to determine whether the read operation was completed using `isDone()`.

Implementing AIO in our Java applications can help us achieve high performance and increased responsiveness and scalability. Next, we will examine buffered input/output.

# Buffered input/output

The buffered approach to input/output, commonly referred to as **buffered input/output**, can be implemented to reduce the number of input/output operations required. This approach is accomplished using temporary storage, known as a **buffer**. Buffers temporarily hold data during the transfer process. The goal of this approach is to minimize direct interactions with hardware and data streams. Java delivers on this promise by accumulating data in a buffer before processing it.

Now that you have a foundational understanding of buffered input/output, let's examine the advantages and best practices of using this technique to improve Java application performance.

## Advantages and best practices

There are several advantages to buffered input/output, and they are best leveraged when following industry best practices. Next, we will review the advantages followed by best practices.

### Advantages

Here are three advantages to implementing buffered input/output in our Java applications:

- **Data handling**: Buffered input/output can increase the efficiency of data handling because it allows for data to be temporarily stored during read and write operations. This is especially beneficial when working with data streams and large files.

- **Flexibility**: We can use buffered classes to encapsulate our input and output streams. This makes them more adaptable for varied uses.

- **Performance**: Buffered input/output aims to reduce the number of input/output operations, thereby reducing interaction overhead and ultimately increasing the overall application performance.

Keeping these advantages in mind, let's review the best practices for optimizing them in Java.

### Best practices

Here are several best practices to help guide your use of buffered input/output:

- **Buffer size**: Testing should be conducted to determine what the most optimal buffer size is. Each use case is different and depends on the data, application, and hardware.

- **Error handling**: It is always good practice to add robust error handling to your applications. This is especially pertinent for situations where input/output operations could fail due to external issues (e.g., file access permissions or network issues).

- **Resource management**: Closing buffered streams will free up system resources and help avoid memory leaks.

We have established that implementing buffered input/output is an approach that can positively impact the performance of our applications. In the next section, we will look at how to implement buffered input/output in Java.

## Implementing buffered input/output

The Java platform includes several classes in the `java.io` library with the following capabilities:

- `BufferedInputStream`: This class is used to read binary data from a stream. The data is stored in a buffer, permitting efficient data retrieval.

- `BufferedOutputStream`: This class writes bytes to a stream, collecting the data in a buffer, and then writes to the output device.

- `BufferedReader`: This class reads from an input stream and buffers the data.

- `BufferedWriter`: This class writes data to an output stream, buffering the data to enable efficient writing operations from the buffer.

Let's look at buffered input/output in code:

```
import java.io.BufferedReader;
import java.io.BufferedWriter;
import java.io.FileReader;
import java.io.FileWriter;
import java.io.IOException;

public class CH12BufferedReadWriteExample {
  public static void main(String[] args) {
    String inputFilePath = "input.txt";
    String outputFilePath = "output.txt";

    try (BufferedReader reader = new BufferedReader(new
    FileReader(inputFilePath));
    BufferedWriter writer = new BufferedWriter(new
    FileWriter(outputFilePath))) {

      String line;
      while ((line = reader.readLine()) != null) {
```

```
        writer.write(line);
        writer.newLine();
      }
   } catch (IOException e) {
      System.err.println("An error occurred: " + e.getMessage());
    }
  }
}
```

As you can see in the example code, we use `BufferedReader` and `BufferedWriter` to read and write to a file. The `readLine` method is used to efficiently read lines from the input file, and `BufferedWriter` can quickly write to a file with minimal input/output operations.

By leveraging the buffered input/output classes covered in this section, and following the provided best practices, we can significantly increase the performance of our Java applications, especially those that have frequent read/write operations.

# Batch operations

The concept of **batch operations** suggests that we process a lot of data at once or combine multiple tasks into a single operation. This type of processing, as opposed to doing things individually, can result in tremendous efficiencies and reduce overhead. Implementing batch operations is usually a good method of improving performance, especially with large-scale data operations, file processing, and database interactions.

Batch operations in practice involve executing a series of jobs, usually with large datasets treated as groups. These groupings are based on natural or logical groupings, with the goal of reducing the computational overhead associated with the repetitive starting and stopping processes.

## Advantages and best practices

There are several advantages to batch operations, and they are best leveraged when following industry best practices. Next, we will review the advantages followed by best practices.

### Advantages

Here are three advantages to implementing batch operations in our Java applications:

- **Performance**: Batch operations represent a tremendous increase in performance. Given the scenario where 100 files need to be processed if handled individually, 100 operations would be required. Handling these 100 files with a batch operation would run faster because there would be a significant reduction in system calls. Network latency would also be improved.

- **Resource usage**: Implementing batch operations reduces overhead and maximizes resource utilization.

- **Scalability**: Batch operations make it easier for our systems to process large datasets. This approach is inherently scalable.

Keeping these advantages in mind, let's review the best practices to optimize them in Java.

### Best practices

Here are several best practices to help guide your use of batch operations:

- **Batch size**: There is a balance between having a batch size that is too small and one that is too big. If it is too small, you are not likely to gain the performance benefits, and if it is too large, your application may run into memory issues. Determining the right size requires testing and is influenced by the type of processing and the type of data involved.

- **Error handling**: As part of your error handling of batch operations, be sure to account for each part of the batch operations if one part fails.

- **Monitoring**: As with all major systems, the importance of logging and monitoring those logs cannot be overstated.

Now, let's see how to implement batch operations in Java.

## Implementing batch operations

The Java platform includes the following APIs and frameworks to help us implement batch operations:

- **Java batch**: The **Java Specification Request** (**JSR**) specification 352 provides a standard approach to batch processing implementation. It includes definitions and steps.

- **JDBC batch processing**: **Java Database Connectivity** (**JDBC**) batch processing is used to handle batch **Structured Query Language** (**SQL**) statements. We will demonstrate this in the following section.

- **Spring batch**: This is a framework that provides a host of batch processing capabilities to include job processing statistics, resource optimization, and transaction management.

Let's look at an example using JDBC batch processing. The following sample program demonstrates how to use JDBC batch processing to efficiently insert multiple records into a database. Note that the database is simulated.

Our example starts with the `import` statements:

```
import java.sql.Connection;
import java.sql.DriverManager;
```

```
import java.sql.PreparedStatement;
import java.sql.SQLException;
```

Next, we create our class and main() method. This section includes connecting to and logging into the simulated database:

```
public class CH12JDBCBatchExample {
  public static void main(String[] args) {
    String url = "jdbc:mysql://localhost/testdb";
    String user = "root";
    String password = "password";
```

The following statement is the SQL statement we plan to use in our batch operation. Following that statement, we use encase our batch operations in a try-catch block:

```
    String sql = "INSERT INTO staff (name, department) VALUES (?, ?)";

    try (Connection conn = DriverManager.getConnection(url, user,
    password);
      PreparedStatement statement = conn.prepareStatement(sql)) {

      conn.setAutoCommit(false); // Turn off auto-commit

      // Add batch operations
      statement.setString(1, "Brenda");
      statement.setString(2, "Marketing");
      statement.addBatch();

      statement.setString(1, "Chris");
      statement.setString(2, "Warehouse");
      statement.addBatch();

      statement.setString(1, "Diana");
      statement.setString(2, "HR");
      statement.addBatch();

      int[] updateCounts = statement.executeBatch();

      conn.commit(); // Commit all the changes

      System.out.println("Rows inserted: " + updateCounts.length);
    } catch (SQLException e) {
```

```
            System.err.println("SQL Exception: " + e.getMessage());
    }
  }
}
```

Our simple example program grouped multiple INSERT statements into a batch, and they were executed in a single request to the simulated database.

Incorporating batch operations in our Java applications, if applicable, can significantly enhance application performance. It can also increase scalability and system maintainability.

# Frameworks for microservices

**Microservices** are a software architectural approach that consists of loosely coupled modules (microservices) that comprise an entire application. The benefits of microservices include the ability to have teams working on individual ones simultaneously, the ability to update one independent of an entire application, increased scalability, and more efficient maintainability. This section focuses on optimization frameworks for microservices.

Now that you have a foundational understanding of frameworks for microservices, let's examine the advantages and best practices of using this technique to improve Java application performance.

## Advantages and best practices

There are several advantages to implementing frameworks for microservices, and they are best leveraged when following industry best practices. Next, we will review the advantages followed by best practices.

### Advantages

Here are three advantages to implementing frameworks for microservices in our Java applications:

- **Fault isolation**: One of the key advantages of using microservices is to isolate faults. This is possible because each microservice is loosely coupled with another one. This means a fault in one microservice will not necessarily impact others.

- **Flexibility**: Adopting the microservice framework provides greater agility and flexibility when developing and maintaining systems.

- **Scalability**: Microservices' distributed nature enables great scalability; they are inherently scalable.

Keeping these advantages in mind, let's review the best practices to optimize them in Java.

## Best practices

Here are several best practices to help guide your use of frameworks for microservices:

- **API design**: When designing and developing APIs, it is important to thoroughly vet their stability and backward compatibility.

- **Configuration management**: A formal version control system should be used so that there is consistency across all microservices.

- **Monitoring**: It is imperative to create a robust logging system and for those logs to be monitored. This can help identify issues before they become critical.

We have established that implementing frameworks for microservices is an approach that can positively impact the performance of our applications. In the next section, we will look at how to implement a framework for microservices in Java.

## Implementing microservices frameworks

Systems designed with a microservice architecture essentially consist of individual services (microservices) that are standalone applications, based on a business function. Each microservice has its own data. There are several frameworks available to implement microservices in Java. Here are four of them:

- **Helidon**: This framework is provided by Oracle and helps us create applications using the microservices architecture. This is a modern framework, and the API offers many options.

- **Micronaut**: This is a modern and robust JVM-based framework that includes features such as dependency injection and container management.

- **Quarkus**: If you use Kubernetes for containerization, Quarkus is a good option to create microservices applications.

- **Spring boot**: This is the most used framework to implement microservices in Java. It is easy to set up, configure, and use.

Let's look at a simple microservice application using Micronaut. We will use a three-step approach:

1. The first step is to set up our project. This can be done with the Micronaut **command-line interface (CLI)** or a supported **Integrated Development Environment (IDE)**, such as IntelliJ IDEA. Using the CLI, here is what the setup code might look like:

   ```
   mn create-app example.micronaut.CH12Service --features=http-
   server
   ```

2.  Next, we need to create a controller to handle HTTP requests. Here is how that can be done:

```
package example.micronaut;

import io.micronaut.http.annotation.Controller;
import io.micronaut.http.annotation.Get;

@Controller("/ch12")
public class CH12Controller {

    @Get("/")
    public String index() {
        return "Hello from CH12's Micronaut!";
    }
}
```

3.  The third and final step is to simply run the application. Note that the following example refers to the **Gradle Wrapper**:

```
./gradlew run
```

Examining the code further, we can see that we use the `@Controller` annotation to identify our class as a controller, whose base URI is `/ch12`. When the application is run, the service will be located at `http://localhost:8080/ch12`.

Microservices frameworks such as Helidon, Micronaut, Quarkus, and Spring Boot provide us with a multitude of capabilities to create Java applications using the microservices architecture.

# Frameworks for cloud-native applications

Developing **cloud-native applications** is a strategic decision, typically based on the desire to exploit the scalability, resiliency, security, and flexibility inherent in cloud computing.

---

**Cloud-native**

In the context of software development, cloud-native refers to the use of cloud computing to create applications from the ground up and deploy them.

---

There are several frameworks available to us that support cloud-native application development in Java. Cloud-native applications are built from start to finish using the cloud environment. These applications are typically built as microservices, which are packaged into containers, orchestrated, and managed using **DevOps**-accepted processes.

Now that you have a foundational understanding of cloud-native applications, let's examine the advantages and best practices of using this technique to improve Java application performance.

# Advantages and best practices

There are several advantages to cloud-native frameworks, and they are best leveraged when following industry best practices. Next, we will review the advantages followed by the best practices.

## Advantages

Here are three advantages to implementing cloud-native frameworks in our Java applications:

- **Efficiencies**: One of the greatest advantages of using cloud-native frameworks for our Java applications is the great efficiencies that are introduced, due to automation, development consistency, and testing.

- **Fault tolerance**: Because cloud-native applications are written as microservices, a fault in one service will not necessarily impact others.

- **Scalability**: The microservices architecture is inherently scalable, as is the cloud environment. This empowers us to build highly scalable Java applications.

Keeping these advantages in mind, let's review the best practices to optimize them in Java.

## Best practices

Here are several best practices to help guide your use of frameworks for cloud-native applications:

- **Containerization**: Applications should be packaged along with their dependencies in containers. This will help ensure that there is consistency with each service, regardless of the runtime environment.

- **Continuous Integration/Continuous Delivery (CI/CD)**: Adopting the CI/CD approach with automated deployment and testing can significantly increase the speed of development and minimize errors inherent in non-automated processes.

- **Monitoring**: Creating robust logs and continuously monitoring them can help identify potential issues before they come to fruition.

We have established that implementing cloud-native frameworks is an approach that can positively impact the performance of our applications. In the next section, we will look at how to implement them in Java.

# Implementing a cloud-native application

There are several frameworks available to us to help develop cloud-native applications in Java. Here are three popular frameworks:

- **Eclipse MicroProfile**: This is a portable API designed for Java enterprise application optimization, specific to microservice architectures. It has many capabilities, including health checks, metrics, and fault tolerance.

- **Quarkus**: This framework uses a **container-first** philosophy and works best with Kubernetes.

- **Spring Cloud**: Part of the Spring environment (i.e., Spring Boot), this is a set of development tools to build common software patterns, such as configurations and service discovery, specific to the cloud environment.

Let's look at a simple cloud-native application using Eclipse MicroProfile:

```
import org.eclipse.microprofile.config.inject.ConfigProperty;
import javax.ws.rs.GET;
import javax.ws.rs.Path;
import javax.ws.rs.Produces;
import javax.ws.rs.core.MediaType;

@Path("/hello")
public class CH12HelloController {

    @ConfigProperty(name = "username")
    String username;

    @GET
    @Produces(MediaType.TEXT_PLAIN)
    public String hello() {
        return "Hello " + username + ", from MicroProfile!";
    }
}
```

The preceding code snippet demonstrates how MicroProfile can be used to inject configuration properties (e.g., a username). This very simple example underscores the benefit of using this type of framework to effectively handle configuration and other microservices-related concerns.

# Case studies and performance analysis

So far in this chapter, we have argued that implementing optimization frameworks can greatly increase the performance and scalability of our Java applications. This section reviews two case studies and explores performance analysis to help determine the impact framework adoption has on our Java applications.

## Case studies

Reviewing case studies based on plausible real-world situations can help demonstrate the advantages of adopting optimization frameworks. Here are two case studies.

## Case study 1

**Name**: A microservices-based e-commerce application

**Background**: The Reinstate LLC company is a large e-commerce retailer that recently transitioned to a microservices architecture from its previous monolithic architecture. They used Spring Boot and Spring Cloud to improve scalability, reliability, and maintainability.

**Opportunity**: Reinstate LLC had difficulty scaling its monolithic system during peak times. They also noted that their development cycles were too long and attributed it to the interconnected nature of their application's components.

**Solution**: Each application component was refactored into a microservice, based on a business function or logic. These microservices included customer profiles, inventory management, order management, and cart management. Spring Boot was used to create individual microservices, and Spring Cloud was used for service discovery, load balancing, and configuration management.

**Result**: After implementing the microservices architecture, Reinstate LLC experienced a 65% reduction in downtime and a 42% improvement in response time.

## Case study 2

**Name**: Financial services' batch processing.

**Background**: CashNow, a financial services company, wanted to process large volumes of transactions each data while ensuring high accuracy and reliability, using batch processing.

**Opportunity**: CashNow's existing system was inefficient and commonly experienced transaction delays. This wreaked havoc on its reporting and end-of-day reconciliation process.

**Solution**: CashNow implemented Spring Batch to help them manage and optimize their batch processing. This framework empowered them to implement job processing, processing chunks, and error handling.

**Result**: CashNow noted a 92% reduction in batch processing time as well as a significant decrease in errors. This change helped them streamline their daily processes to include end-of-day reconciliation and reporting.

## Performance analysis

Implementing optimization frameworks is only one part of the solution. Once they are implemented, we want to ensure they result in improvements. We also want to make sure the changes work the way we intend them to. This is typically accomplished by observing metrics over time.

A common performance analysis approach involves profiling tools (see *Chapter 14, Profiling Tools*) and monitoring tools. Examples include **JProfiler** and **VisualVM**. Using robust profiling and monitoring tools can help us identify potential bottlenecks, such as memory leaks and slow database queries.

The case studies and performance analysis presented in this section underscore the importance of implementing optimization frameworks in our Java applications to help increase performance, scalability, and maintainability.

## Summary

This chapter explored various optimization frameworks and techniques to help increase the performance of our Java applications, support scalability, make our code more maintainable, and provide efficient development cycles. We covered asynchronous input/output, buffered input/output, batch operations, frameworks for microservices, and frameworks for cloud-native applications. We concluded the chapter with a review of two realistic case studies and an overview of performance analysis.

Hopefully, the comprehensive overview provided in this chapter will help you further optimize your Java applications. The frameworks and techniques covered in this chapter can help you enhance the performance of your applications and increase scalability, consistency, reliability, and maintainability.

In the next chapter (*Chapter 13, Performance-Focused Libraries*), we will explore several open source Java libraries designed to provide high performance. These fully optimized libraries can be leveraged to our advantage. Notable libraries covered include **Java Microbenchmark Harness** (**JHM**), which is part of the OpenJDK project; Netty to work with network protocols, which can be used to reduce latency; and FasterXML Jackson, which is a suite of data processing tools.

# 13

# Performance-Focused Libraries

The performance of modern Java applications is a paramount concern that can significantly impact the success of an application and the organization. Performance can include execution speed, network responsiveness, and data handling optimization. Regardless of the type of performance you are trying to improve, selecting and properly implementing the most ideal tools and libraries are key to bringing your performance improvement goals to fruition.

This chapter highlights a specific set of tools that can be instrumental in improving the performance of Java applications. The first tool reviewed is **Java Microbenchmark Harness (JMH)**, which helps us create reliable benchmarks. Our JMH coverage will include fundamental knowledge and hands-on application. Netty, a network application framework focused on high performance, will also be covered. This framework's greatest value is in applications that require rapid response times or scalable network architectures.

Our coverage of performance-focused libraries includes an examination of FasterXML Jackson, a high-speed **JavaScript Object Notation (JSON)** processor that, as you will have the opportunity to learn, facilitates data processing efficiencies. FasterXML Jackson, also referred to as just Jackson, has streaming and data-binding APIs that can significantly improve performance when working with JSON data. The chapter concludes with a section on other notable libraries, including Eclipse Collections, Agrona, and Guava.

By the end of this chapter, you should have a foundational understanding of performance-focused libraries and be able to leverage the knowledge you gained from the hands-on exercises to improve the performance of your Java applications.

This chapter covers the following main topics:

- Java Microbenchmark Harness
- Netty
- FasterXML Jackson
- Other notable libraries

# Technical requirements

To follow the examples and instructions in this chapter, you will need the ability to load, edit, and run Java code. If you have not set up your development environment, refer to *Chapter 1, Peeking Inside the Java Virtual Machine (JVM)*.

The finished code for this chapter can be found here: `https://github.com/PacktPublishing/High-Performance-with-Java/tree/main/Chapter13`.

# Java Microbenchmark Harness

Benchmarking is critical to the ability to measure performance. JMH is a toolkit used to implement rigorous benchmarks. This section explores JMH and its key features and provides implementation examples.

> **Java Microbenchmark Harness (JMH)**
>
> JMH is used to build and implement benchmarks to analyze the performance of Java code. It was written in Java by the team that created the **Java Virtual Machine (JVM)**.

Developers that use JMH can measure the performance of Java code snippets with repeatable and controlled conditions.

## Key features

JMH is an open source toolkit used to build and implement benchmarks at the nano, micro, and macro levels. JMH is more than a performance tester; it is designed to overcome or avoid common performance measurement pitfalls, including warm-up times and the effects of **just-in-time** (**JIT**) compilations.

Key features of the JMH toolkit include the following:

- **Annotations**: As you will see in the next section, JMH uses Java annotations to easily define benchmarks. This feature is developer-friendly.

- **JVM integration**: JMH works in step with JVM intervals. This provides us with consistent and reliable results.

- **Microbenchmarking support**: JMH, as the name suggests, focuses on small code snippets. This helps increase the accuracy of performance measurements.

Now that you have a basic understanding of the JMH toolkit, let's look at how to write benchmarks.

## Adding JMH libraries to your IDE

Depending on your **Integrated Development Environment** (**IDE**) setup, you may have to manually add the JMH libraries to your Java project. The following steps can be used with Visual Studio Code to add the JMH libraries to a Java project:

1.  Create a **Maven quickstart** project.

2.  Edit the pom.xml file by adding the following dependencies:

    ```
    <dependency>
        <groupId>org.openjdk.jmh</groupId>
        <artifactId>jmh-core</artifactId>
        <version>1.32</version>
        <scope>test</scope>
    </dependency>
    <dependency>
        <groupId>org.openjdk.jmh</groupId>
        <artifactId>jmh-generator-annprocess</artifactId>
        <version>1.32</version>
        <scope>test</scope>
    </dependency>
    ```

3.  Save and reload the project using **Maven: Reload project** via the command palette.

You might opt to use a more robust IDE, such as IntelliJ IDEA. Here are the steps to add these libraries to a project in that IDE:

1.  Create a project in IntelliJ IDEA.

2.  Select the **File | Project Structure** menu option.

3.  In the **Project Structure** interface, select **Project Settings | Libraries**.

4.  Click the + button to add a new library.

5.  Choose the **From Maven** option.

6.  Use the search feature to find the most recent version of org.openjdk.jmh:jmh-core and click **OK** to add the library to your project.

7.  Use the search feature to find the most recent version of org.openjdk.jmh:jmh-generator-annprocess and click **OK** to add the library to your project.

8.  Click **Apply** to apply the changes and then **OK** to close the **Project Structure** dialog window.

9.  Lastly, ensure that the libraries have been automatically added to your module. If this is not the case, select the **File | Project Structure | Modules** menu option. If the new JMH libraries are not listed in the **Dependencies** area, use the + button to add them.

If you are not using Visual Studio Code or IntelliJ IDEA, follow the steps appropriate for your IDE.

# Writing benchmarks

To use JMH in Java code, we simply add the `@Benchmark` annotation and use JMH's core APIs to configure and execute our benchmarks. Let's look at an example in code. Our example tests two methods, both with a different approach to string reversal:

```java
package org.example;

import org.openjdk.jmh.annotations.Benchmark;
import org.openjdk.jmh.annotations.Setup;
import org.openjdk.jmh.annotations.State;
import org.openjdk.jmh.annotations.Scope;
import org.openjdk.jmh.runner.Runner;
import org.openjdk.jmh.runner.options.Options;
import org.openjdk.jmh.runner.options.OptionsBuilder;

@State(Scope.Thread)
public class Ch13StringBenchmark {

    private String sampleString;

    @Setup
    public void prepare() {
        sampleString = "The quick brown fox jumps over the lazy dog";
    }

    @Benchmark
    public String reverseWithStringBuilder() {
        return new StringBuilder(sampleString).reverse().toString();
    }

    @Benchmark
    public String reverseManually() {
        char[] strArray = sampleString.toCharArray();
        int left = 0;
        int right = strArray.length - 1;
        while (left < right) {
            char temp = strArray[left];
            strArray[left] = strArray[right];
            strArray[right] = temp;
            left++;
            right--;
        }
        return new String(strArray);
```

```
    }

    public static void main(String[] args) throws Exception {
        Options opt = new OptionsBuilder()
                .include(Ch13StringBenchmark.class.getSimpleName())
                .forks(1)
                .build();

        new Runner(opt).run();
    }
}
```

Next, let's run our code and review the results.

## Running benchmarks

Once we define our benchmarks, we simply run them via the `main()` method in our Java application or, alternatively, via the command line. If our application does not contain a `main()` method, then we would run the class containing the benchmarks. In our previous code example, that would be the `Ch13StringBenchmark()` class. JMH provides detailed output with time measurements and throughput rates. Analysis of this data can provide significant performance-related insights into the benchmarked code.

Even our simple test results in extensive output. The final segment of the output is provided in the following figure. The complete output is provided in the `Ch13StringBenchmarkOutput.txt` file in this chapter's GitHub repository.

| Benchmark | Mode | Cnt | Score | Error | Units |
|---|---|---|---|---|---|
| Ch13StringBenchmark.reverseManually | thrpt | 5 | 14455985.146 ± 1076836.292 | | ops/s |
| Ch13StringBenchmark.reverseWithStringBuilder | thrpt | 5 | 28375777.868 ± 2268015.114 | | ops/s |
| Class transformation time: 0.011536673s for 923 classes or 1.249910400866739E-5s per class | | | | | |

Figure 13.1 – Final benchmark output

Referring to the preceding output, let's look at how that information can be analyzed to provide performance insights.

## Analyzing results

As you saw in the previous section, JMH provides extensive output. Looking at the final three lines of output illustrated in *Figure 13.1*, there are several columns that we should understand:

- `Mode`: This is the benchmark mode. Our mode was `thrpt` for throughput. It could have alternatively been `avgt` for average time.

- `Cnt`: This is the count of benchmark iterations. In our case, it was 5 for each benchmark.

- `Score`: This is the benchmark score that shows, in our case, the average throughput time in microseconds.

- `Error`: This column contains the margin of error for the score.

Based on the output, we can see that the first benchmark was faster than the second benchmark. Viewing these types of results can help developers decide how to implement certain functionality to achieve high performance in their code.

## Use cases

There are a few common JMH use cases:

- Algorithm optimization

- Comparative analysis

- Performance regression tests

JMH empowers Java developers to create and implement benchmarks. Analyzing the results can help developers make informed decisions based on an analysis of empirical data. This can lead to more performant Java applications.

# Netty

Netty, cutely named for networking, is a high-performance, event-driven application framework. This framework helps developers create network applications by simplifying network functionality programming such as with **User Datagram Protocol** (**UDP**) and **Transmission Control Protocol** (**TCP**) socket servers.

Network programming often involves low-level APIs and Netty provides a level of abstraction making it easier to develop with. The Netty architecture is scalable, supports many connections, and is designed to minimize latency and resource overhead.

## Core features

Netty is the framework of choice for many network developers due to its reliability, scalability, and ease of use. The core features of Netty include the following:

- **Built-in codec support**: Netty has built-in encoders and decoders to help developers work with various protocols including HTTP and WebSocket. Netty negates the need for separate implementations for supported protocols.

- **Customizable pipeline**: The Netty framework includes a pipeline architecture that facilitates data encapsulation and handlers. It uses a modular approach, making pipeline configuration an easy task for developers.

- **Event-driven**: Netty's event-driven design results in asynchronous input/output handling. This non-blocking approach minimizes network latency.

Armed with an understanding of Netty's core features, let's review performance considerations.

## Performance considerations

Our focus throughout this book has been on high-performance Java applications. Netty is a great addition to our high-performance tool kit. It emphasizes high performance with its **thread model flexibility** and **zero-copy capabilities**. Let's look at those performance considerations:

- **Thread model flexibility**: Netty's thread management is highly configurable. Developers can configure Netty to manage their application's threads based on specific use cases, such as scaling up or down and reducing the number of threads.

- **Zero-copy capabilities**: Netty's zero-copy API helps make data processing (input and output) more efficient. This is accomplished by minimizing unnecessary memory duplication.

Let's look at an example of using Netty to create an echo server that simply echoes data it receives to the client.

## Implementation

The following example demonstrates the relative ease with which Netty can be used to handle network events and how Netty can facilitate high performance in network communications. Note that you will document your dependencies in the pom.xml file:

```
import io.netty.bootstrap.ServerBootstrap;
import io.netty.channel.*;
import io.netty.channel.nio.NioEventLoopGroup;
import io.netty.channel.socket.SocketChannel;
import io.netty.channel.socket.nio.NioServerSocketChannel;
import io.netty.handler.codec.string.StringDecoder;
import io.netty.handler.codec.string.StringEncoder;

public class Ch13EchoServer {
  public static void main(String[] args) throws InterruptedException {
    EventLoopGroup bossGroup = new NioEventLoopGroup(1);
    EventLoopGroup workerGroup = new NioEventLoopGroup();
    try {
```

```
        ServerBootstrap b = new ServerBootstrap();
        b.group(bossGroup, workerGroup)
         .channel(NioServerSocketChannel.class)
         .childHandler(new ChannelInitializer<SocketChannel>() {
         @Override
         protected void initChannel(SocketChannel ch) throws Exception
{
            ChannelPipeline p = ch.pipeline();
            p.addLast(new StringDecoder(), new StringEncoder(), new
            Ch13EchoServerHandler());
         }
        });

        ChannelFuture f = b.bind(8080).sync();
        f.channel().closeFuture().sync();
    } finally {
        workerGroup.shutdownGracefully();
        bossGroup.shutdownGracefully();
    }
}

static class Ch13EchoServerHandler extends
ChannelInboundHandlerAdapter {
   @Override
   public void channelRead(ChannelHandlerContext ctx, Object msg) {
     ctx.write(msg);
     ctx.flush();
   }
  }
}
```

Here is simulated output of the Netty echo server:

```
Server started on port 8080.
Client connected from IP: 192.168.1.5
Received from client: Hello Server!
Echoed back: Hello Server!
Received from client: This is a test message.
Echoed back: This is a test message.
Received from client: Netty is awesome!
Echoed back: Netty is awesome!
Client disconnected: IP 192.168.1.5
Server shutting down...
```

Netty is a mature and robust framework for network application development. It is inherently scalable and offers performance benefits with network functionality. Netty also introduces development efficiencies and shorter development times.

# FasterXML Jackson

JSON is a file format used for interchanging data. It is structured and human-readable text used to transmit data objects. The format consists of arrays and attribute-value pairs. The example JSON object provided in the following code block represents a user profile for a social media system. As you can see, the fields contain the user's name, age, email address, and hobbies:

```
{
    "name": "Neo Anderson",
    "age": 24,
    "email": "neo.anderson@matrix.com",
    "hobbies": ["coding", "hacking", "sleeping"]
}
```

This JSON object consists of attribute-value or key-value pairs and an array of strings for the user's hobbies. JSON is a common method of data representation in data storage and web applications.

**FasterXML Jackson** is a library with the primary ability to rapidly process and create JSON objects. These objects are read sequentially and FasterXML Jackson, referred to from this point as **Jackson**, uses a cursor to keep track of its place. Jackson is lauded as a performance maximizer and memory minimizer.

The *XML* in *FasterXML Jackson* suggests it can handle XML files too. In addition to Jackon's ability to rapidly process JSON, it can also process **Comma-Separated Values (CSV)**, **eXtensible Markup Language (XML)**, **YAML Ain't Markup Language (YAML)**, and other file formats.

## Core features

Core features of Jackson include the following:

- **Data binding**: Data binding is a Jackson feature that supports efficient and reliable conversion between Java objects and JSON text. Implementation is straightforward.

> **Data binding**
> A technique in computer programming that links (binds) data sources to the sender (provider) and receiver (client).

- **Streaming API**: Jackon has a highly efficient, low-level streaming API for parsing and generating JSON.

- **Tree model**: When flexible JSON operations are needed, Jackson's tree model can be implemented to represent JSON documents in a tree structure – a tree of nodes. This is often used when the JSON structures are complex.

Now that you understand Jackson's core features, let's review performance considerations.

## Performance considerations

The performance considerations detailed here illustrate how Jackson was designed as a performance-focused library:

- **Custom serialization and deserialization**: Jackson empowers developers to define their own serializers and deserializers for custom fields. This can lead to significant performance enhancements.

- **Zero-copy**: Like Netty, Jackson's zero-copy API helps make data processing (input and output) efficient. This is accomplished by minimizing unnecessary memory duplication.

Let's look at an example of using Jackson to serialize and deserialize Java objects.

## Implementation

Let's start by adding dependencies to our `pom.xml` file to include Jackson in our project. Here is what that might look like:

```
<dependencies>
    <dependency>
        <groupId>com.fasterxml.jackson.core</groupId>
        <artifactId>jackson-core</artifactId>
        <version>2.12.3</version>
    </dependency>
    <dependency>
        <groupId>com.fasterxml.jackson.core</groupId>
        <artifactId>jackson-databind</artifactId>
        <version>2.12.3</version>
    </dependency>
</dependencies>
```

The following example illustrates how straightforward Jackson is for developing object serialization and deserialization:

```
import com.fasterxml.jackson.databind.ObjectMapper;

public class Ch13JacksonExample {
    public static void main(String[] args) throws Exception {
```

```
    ObjectMapper mapper = new ObjectMapper();

    // Example of a Plan Old Java Object (POJO)
    class User {
      public String name;
      public int age;

      // Constructors, getters, and setters have been omitted for
      // brevity
    }

  // Serialize Java object to JSON
  User user = new User();
  user.name = "Neo Anderson";
  user.age = 24;
  String jsonOutput = mapper.writeValueAsString(user);
  System.out.println("Serialized JSON: " + jsonOutput);

  // Deserialize JSON to Java object
  String jsonInput = "{\"name\":\"Neo Anderson\",\"age\":24}";
  User userDeserialized = mapper.readValue(jsonInput, User.class);
  System.out.println("Deserialized user: " + userDeserialized.name);
  }
}
```

In this section, we learned that Jackson is a key tool for processing JSON. It is fast, flexible, and robust. Jackson can be the first tool you think of when needing to work with JSON in your Java applications.

# Other notable libraries

So far, we have covered JMH, Netty, and Jackson and posited that they are core libraries focused on high performance in Java. Each is designed for a specific type of task. There are other libraries available that are worth learning about. This section explores three additional libraries: **Agrona**, **Eclipse Collections**, and **Guava**.

## Agrona

Agrona is a data structure collection specifically designed for creating high-performance Java applications. These data structures include maps and ring buffers. An example use case is a stock trading application whose success hinges on low latency.

Key features of Agrona include the following:

- Non-blocking data structures – this supports high throughput and results in low latency
- Specifically designed for high-frequency stock and securities trading systems
- Uses direct buffers, contributing to off-heap memory management efficiencies

Let's look at an implementation example of Agrona that illustrates how to use a specific high-performance data structure. For our example, we will use the ManyToOneConcurrentArrayQueue data structure:

```java
import org.agrona.concurrent.ManyToOneConcurrentArrayQueue;

public class Ch13AgronaExample {
  public static void main(String[] args) {
    // Create a queue with a capacity of 10 items
    ManyToOneConcurrentArrayQueue<String> queue = new
    ManyToOneConcurrentArrayQueue<>(10);

    // Producer thread that offers elements to the queue
    Thread producer = new Thread(() -> {
      for (int i = 1; i <= 5; i++) {
        String element = "Element " + i;
        while (!queue.offer(element)) {
          // Retry until the element is successfully added
          System.out.println("Queue full, retrying to add: " +
          element);
          try {
            Thread.sleep(10); // Sleep to simulate backoff
          } catch (InterruptedException e) {
            Thread.currentThread().interrupt();
          }
        }
        System.out.println("Produced: " + element);
      }
    });

    // Consumer thread that polls elements from the queue
    Thread consumer = new Thread(() -> {
      for (int i = 1; i <= 5; i++) {
        String element;
        while ((element = queue.poll()) == null) {
          // Wait until an element is available
          System.out.println("Queue empty, waiting for elements...");
          try {
```

```
            Thread.sleep(10); // Sleep to simulate processing delay
        } catch (InterruptedException e) {
            Thread.currentThread().interrupt();
        }
    }
    System.out.println("Consumed: " + element);
    }
});

// Start both threads
producer.start();
consumer.start();

// Wait for both threads to finish execution
try {
    producer.join();
    consumer.join();
} catch (InterruptedException e) {
    Thread.currentThread().interrupt();
    }
  }
}
```

As you can see in the preceding code and surmise from the in-code comments, we initialize `ManyToOneConcurrentArrayQueue` with a capacity value of `10`. This type of queue is well suited for use cases where there is a single **customer** and multiple **producers**. Our example includes consumer and producer threads. The code implemented basic thread handling.

## Eclipse Collections

Eclipse Collections is a set of memory-efficient algorithms and data structures. These collections can be used to significantly improve performance. Engineered for large-scale systems, Eclipse collections come in both mutable and immutable forms. They offer efficient memory management.

Key features of Eclipse collections include the following:

- A comprehensive set of data structures to include bags, lists, maps, sets, stacks, and more

- Primitive collections and associated classes

- Utility methods that can transform collections and be used for filtering, iterating, and sorting

Let's demonstrate how to use an `ImmutableList` from an Eclipse collection. This is one of the more memory-efficient collections:

```java
import org.eclipse.collections.api.factory.Lists;
import org.eclipse.collections.api.list.ImmutableList;

public class Ch13EclipseCollectionsExample {
    public static void main(String[] args) {
        // Creating an immutable list using Eclipse Collections
        ImmutableList<String> immutableList = Lists.immutable.
        of("Apple", "Pear", "Cherry", "Lime");

        // Displaying the original list
        System.out.println("Original immutable list: " +
        immutableList);

        // Adding an item to the list, which returns a new immutable
        // list
        ImmutableList<String> extendedList = immutableList.
        newWith("Orange");

        // Displaying the new list
        System.out.println("Extended immutable list: " +
        extendedList);

        // Iterating over the list to print each element
        extendedList.forEach(System.out::println);
    }
}
```

Our example starts with the creation of an immutable list of fruits. Next, we add an element to the list, then we iterate through the list for output.

To use Eclipse Collections in our applications, we need to include the library in our project. With the example of Maven, we would simply add the following to our pom.xml file:

```xml
<dependency>
    <groupId>org.eclipse.collections</groupId>
    <artifactId>eclipse-collections-api</artifactId>
    <version>11.0.0</version>
</dependency>
<dependency>
    <groupId>org.eclipse.collections</groupId>
    <artifactId>eclipse-collections</artifactId>
    <version>11.0.0</version>
</dependency>
```

Our code snippet provided a basic introduction to using Eclipse Collections.

## Guava

Guava is a product from Google that includes new collection types such as multimap and multiset. It also includes immutable collections, a graph library, support for primitives, and caching utilities. Here is a list of key features of Guava:

- Advanced collection types.

- Advanced collection utilities.

- Caching support using CacheBuilder. This can be used to improve application speed.

- Utilities for concurrency.

- Utilities for hashing.

- Utilities for input/output operations.

Here is an example application that demonstrates the use of Guava's CacheBuilder. The application creates a cache that automatically loads and stores value-based keys:

```
import com.google.common.cache.CacheBuilder;
import com.google.common.cache.CacheLoader;
import com.google.common.cache.LoadingCache;
import java.util.concurrent.TimeUnit;

public class Ch13GuavaExample {
  public static void main(String[] args) throws Exception {
    LoadingCache<String, String> cache = CacheBuilder.newBuilder()
      .maximumSize(100)
      .expireAfterWrite(10, TimeUnit.MINUTES)
      .build(
        new CacheLoader<String, String>() {
          public String load(String key) {
            return "Value for " + key;
          }
        }
      );

    System.out.println(cache.get("key1"));
    System.out.println(cache.get("key2"));
    System.out.println(cache.get("key3"));
  }
}
```

Java has several libraries and frameworks that can be used to improve overall performance. Understanding what libraries are available to us and how to implement them can be crucial to an application's success.

## Summary

This chapter explored several key high-performance libraries that we can employ to improve the performance of our Java applications. Specifically, we reviewed JMH and indicated that it provides reliable performance benchmarking. We also looked at Netty and identified its applicability to improving the performance of network applications. FastXML Jackson was also reviewed for its specialized use in handling JSON objects. Lastly, we covered three additional libraries: Agrona, Eclipse Collections, and Guava.

Each library featured in the chapter is tailored to a specific Java programming need. These tools are poised to help us significantly improve the performance of our Java applications. Experimenting with these libraries with your own Java projects can help solidify your understanding of them and the best use case for each of them. Furthermore, implementing these libraries appropriately can lead to overall improved performance of your Java application.

# 14
# Profiling Tools

As the complexity of our Java applications increases, the need to gain insights into how they use system resources such as CPU and memory becomes increasingly important and a critical aspect of ensuring our applications perform efficiently. This is where profiling tools come in; they can help us identify issues such as bottlenecks and memory leaks so that we can enhance our applications to improve the user experience and overall performance.

This chapter takes a deep dive into profiling and profiling tools. We will start with an introduction to profiling and its importance to our ability to fine-tune our applications for optimal performance. Categories of profiling tools and their uses are also covered to help give you a basic understanding, leading to a review of specific profiling tools.

We will cover Java profiling tools bundled with the **Java Development Kit** (**JDK**) and ones embedded in **Integrated Development Environments** (**IDEs**), such as IntelliJ IDEA, Eclipse, and NetBeans. Additionally, third-party profiling tools will be reviewed, including YourKit Java Profiler, JProfiler, and VisualVM. The intention is to provide you with a firm understanding of the various profiling tools available, their strengths and weaknesses, and practical use cases so that you can determine which tool is best suited for your needs and use them effectively.

Our coverage of profiling tools includes a comparative analysis to help you evaluate performance overhead, tool accuracy, ease of use, integration issues, and costs. We will end the chapter with a look at future Java profiling trends, including profiling tool advancements, emerging standards, and the integration of **Artificial Intelligence** (**AI**) and **Machine Learning** (**ML**) for further performance tuning.

By the end of this chapter, you should have a foundational understanding of profiling tools and be able to leverage the knowledge you gained from hands-on exercises to improve the performance of your Java applications.

This chapter covers the following main topics:

- An introduction to profiling tools
- Profilers bundled in the JDK
- IDE-embedded profilers

- Additional profilers

- A comparative analysis of profiling tools

- Practical profiling strategies

- Case studies

- Future trends in Java profiling

# An introduction to profiling tools

Profiling tools play a critical role in performance tuning and proactive application enhancements to support high-performance Java applications.

> **Profiling**
>
> Profiling is the process of analyzing software at runtime with the goal of identifying performance issues, including bottlenecks, resource use, and optimization opportunities.

Once we build and test our software, it moves into production. This is where profiling occurs, during runtime. Using profiling tools, we can gain detailed insights into the runtime behavior of our applications. The goal is to have efficient code and our systems to perform optimally, including low latency, high reliability, and high availability. It is not enough for our applications to perform accurately; they also need to perform efficiently. The use of profiling tools allows us to pinpoint performance issues so we can further optimize our code.

## The importance of profiling in performance tuning

Let's look at five specific reasons that illustrate the importance of profiling to support performance tuning:

- **Application responsiveness**: At the heart of performance tuning is ensuring that our applications are responsive. Users should not be subjected to high latency. Profiling tools help us analyze how our resources are used, shedding light on opportunities for improvement.

- **Bottlenecks**: The use of profiling tools can help us identify performance bottlenecks and even potential bottlenecks. This allows us to take a proactive approach to performance tuning and helps us avoid catastrophic bottlenecks in the future.

- **Continuous improvement**: Software systems are not developed, deployed, and forgotten. We maintain our systems and strive to continually improve them. Systems that perform at desired levels today might underperform when a new operating system is released, or other environmental factors change. This requires a continuous improvement mindset. Continually profiling our applications and addressing optimization opportunities can help ensure our systems remain responsive and efficient.

- **Resource utilization**: The use of hardware resources such as CPU and memory should be closely monitored, as inefficient use of these resources can result in system lag and suboptimal input/output operations. Profiling tools can help us identify areas that can be optimized.

- **Scalability**: The larger our applications are, the more significant even a minor performance issue can be to overall system performance. The use of profiling tools helps us identify opportunities to enhance performance and address any related issues that could become more pronounced as our application scales, due to increased demand.

## Types of profiling tools and their uses

There are seven broad categories of profiling tools, each with its own focus area or purpose. Let's review those now:

- **CPU profilers**: This category of profiles analyzes an application's CPU usage. They can identify specific methods in our applications that are the most CPU-intensive.

- **IDE embedded profilers**: Most major IDEs (e.g., NetBeans, Eclipse, and IntelliJ IDEA) have built-in profiling tools. Because they are part of the IDE, they seamlessly integrate with the development environment. This makes them especially easy to use.

- **Input/output profilers**: These profilers specifically focus on our application's input/output operations. If our input/output processes are slow or inefficient, these profilers can bring them to light so that we can refine operations such as file handling, database interactions, and network communication.

- **Memory profilers**: Memory profilers give us a window into memory allocation and memory use. This can help us identify actual or potential memory leaks and identify memory consumption optimization opportunities. We can use this category of profiler to provide insights into garbage collection, memory retention, and even object life cycles.

- **Network profilers**: As the name suggests, this category of profilers specializes in analyzing an application's network communications. These profilers can help us identify latency issues and to understand how bandwidth is being used. Network profilers also help us identify suboptimal or inefficient network protocols.

- **Specialized profilers**: This is a category of profilers that provide tailored features for specific performance situations. They can also be used for specific environments (e.g., distributed systems or real-time systems) or specific scenarios.

- **Thread profilers**: Thread profilers monitor thread activity and are especially useful in multithreaded environments. This category of profiler helps us detect thread contention issues and potential deadlocks and identifies opportunities to optimize thread management.

It is important to understand what profiling tools are and their categories and uses. This knowledge can help you select the most appropriate tool for your performance-related goals and system requirements.

# Profilers bundled in the JDK

The JDK includes profiling tools that can provide us with valuable insights into our Java application's performance. The use of these tools is highly recommended to identify performance issues and opportunities for performance fine-tuning. This section explores the primary profilers that are built-in to the JDK – JVisualVM and **Java Mission Control (JMC)**.

## JVisualVM

The **JVisualVM**, short for **Java Visual Virtual Machine**, is a robust profiling tool built into the JDK. It offers an impressive set of features that include monitoring, profiling, and troubleshooting Java applications.

### An overview and features

JVisualVM provides a graphical user interface that combines several JDK tools, including JConsole. The interface is displayed in the following figure.

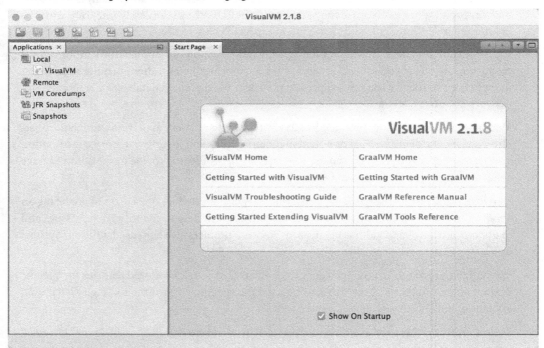

Figure 14.1: The JVisualVM graphical user interface

The key features of JVisualVM are summarized as follows:

- **Heap dump analysis**: JVisualVM is able to capture heap dumps and analyze them for us. This can help us gain insight into memory use and identify real or potential memory leaks.

- **Plugin integration**: This tool's functionality can be extended by a multitude of available plugins. The plugin repository is available via the Tools menu, open and displayed in the following figure.

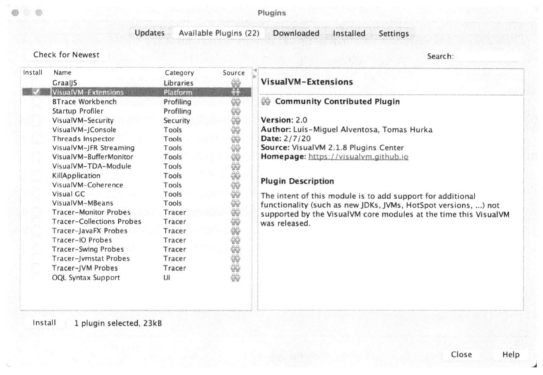

Figure 14.2: The JVisualVM plugin repository

- **Monitoring**: JVisualVM can monitor, in real time, our application's memory consumption, garbage collection, thread activity, and CPU usage.

- **Profiling**: This tool offers the ability to conduct detailed profiling of CPU and memory use. This can help identify memory leaks and bottlenecks.

- **Thread analysis**: JVisualVM can be used to monitor and analyze our application's thread states and thread activities.

## *Use cases and examples*

We can use JVisualVM for multiple use cases, and it is especially useful in the development, testing, and production phases. This tool provides us with detailed CPU and memory usage insights. We can use this tool to see which of our methods consumes the most CPU time. We can also look at memory consumption, object allocations, and garbage collection to help us optimize our application's memory use.

As an example, consider a situation where we have a web application whose users report occasional lag. We can use JVisualVM to monitor our application's CPU and memory use. This can help us identify spikes. From there, we can analyze thread dumps to determine the source of the problem. In this scenario, it could be as simple as a single Java method or thread that causes the occasional lag. Using the profiling tool can help us quickly get to the core of the issue so that we can refine our code accordingly.

## JMC

JMC is another powerful JDK tool that we can use to profile and monitor our Java applications, most specifically in production environments. As you can see from the screenshot below, JMC includes the **JMX Console** and **Flight Recorder**.

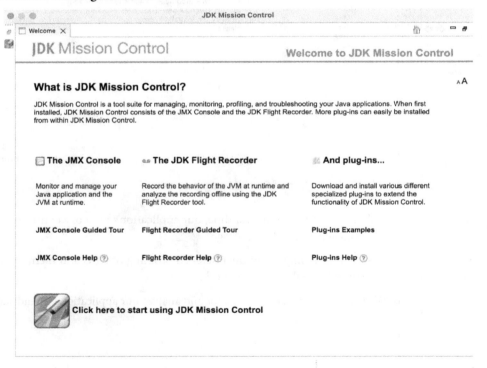

Figure 14.3: The JMC welcome screen

### An overview and features

The **Java Management Extensions** (**JMX**) console is used to monitor and manage our Java applications. The **Flight Recorder** is for continuous monitoring and profiling. It collects detailed performance data, with limited impact on application performance. The events recorded include method invocations, memory allocation, thread activity, and input/output operations. The **Java Flight Recorder** (**JFR**) is the core component of JMC and can be used to record our running applications and then analyze the results, giving us insights into CPU usage, method-specific execution times, data on memory allocations, and so on.

Leveraging the capabilities of JMC and JVisualVM can result in deep insights into our application's performance, empowering us to optimize our resource use and improve the responsiveness of our applications. In the next section, we will review profiling tools embedded in IDEs.

# IDE-embedded profilers

Java developers use built-in profiling tools with their favorite IDEs. Using these profilers offers a convenient method of analyzing our software directly within our development environment of choice. This section explores the built-in capabilities of the IntelliJ IDEA, Eclipse, and NetBeans IDEs.

## The IntelliJ IDEA profiler

The **IntelliJ IDEA profiler** allows us to profile our Java applications from within the IDE. This powerful feature is only available in the commercial edition of the IDE (**IntelliJ IDEA Ultimate**), so if you are using the **Community Edition** (**CE**), you will not be able to use the IntelliJ IDEA profiler. Integration and setup are simple; here are the steps:

1.  Open IntelliJ IDEA.
2.  Write or load your Java project.
3.  Use the **Run | Profile** menu option to start profiling.
4.  Select the type of profiling you want (CPU or memory).

With CPU profiling, the tool can identify methods in your code that consume the most CPU time. Call trees are also provided that show code paths and overall resource use (CPU and memory). The memory profiling capabilities of the IntelliJ IDEA Profiler include the ability to trace object allocations, analyze garbage collection efficacy, and detect memory leaks.

## Eclipse's Test and Performance Tools Platform

The Eclipse IDE is popular for Java developers, and it previously had an embedded profiling tool called the **Test and Performance Tools Platform** (**TPTP**). It was provided as a plugin, available from the Eclipse Marketplace. The key features included are as follows:

- CPU profiling
- Memory profiling
- Thread analysis
- Input/output profiling
- Network activity

TPTP is mentioned here so that developers with Eclipse are not left wondering what embedded profiling tools Eclipse has. TPTP was archived, likely due to the increasing power of the JDK built-in profiling tools.

## NetBeans Profiler

The NetBeans Profiler is integrated into the NetBeans IDE and has a top-level **Profile** menu item for easy configuration and access. The tool can conduct performance analysis in the categories of telemetry, methods, objects, threads, locks, and SQL queries.

The following screenshot shows the four main components of the tool's dashboard when **Telemetry profiling** is selected.

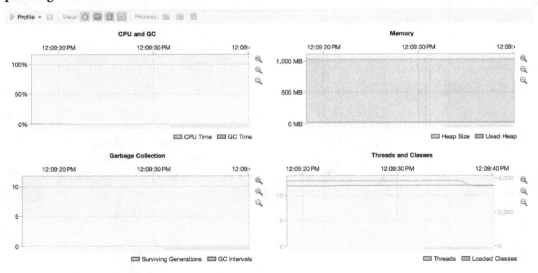

Figure 14.4: The NetBeans telemetry profiler window

The object profiler, shown in the following figure, provides real-time analysis so that you can review your application's performance at runtime.

Figure 14.5: The NetBeans object profiler window

Understanding the profiling tools that are integrated with your IDE can help you become more efficient with your time, as these integrated tools are the easiest to configure and use when compared to third-party tools, and in some cases, they are easier than using the profiling tools bundled with the JDK.

# Additional profilers

So far, we have looked at profiling tools that are built-in to the JDK and ones embedded in an IDE. There is a third category of profilers, ones that are external to the JDK and IDEs. This section reviews three of the most prevalent profilers in this category. These are specialized tools that can give us deeper insights, based on sophisticated analysis of our runtime applications.

## YourKit Java Profiler

YourKit Javak Profiler is a robust profiling tool that is available with an open source license. Advanced features and support require purchasing a license. The key features and capabilities of this tool are listed as follows:

- **Continuous Integration/Continuous Delivery (CI/CD)** integration
- CPU profiling
- Database profiling
- Input/output profiling
- Memory profiling
- Remote server profiling
- Thread profiling

## JProfiler

The JProfiler profiling tool is a commercial application that is said to be an intuitive, all-in-one profiler for Java applications. It has a friendly, easy-to-use interface and powerful capabilities that include the following:

- CPU profiling
- Heap and object graph analysis
- IDE integration
- Memory profiling
- Remove server profiling
- SQL profiling
- Thread profiling

## VisualVM plugins and extensions

We previously covered VisualVM as a JDK-bundled profiling tool. It is featured in this section because there are a host of third-party plugins and extensions that can be used to enhance VisualVM.

The following list shows some of the plugins available:

- **Buffer Monitor**: Can be used to monitor direct buffer use.
- **Heap Walker**: This plugin provides finite memory and heap analysis.
- **Kill Application**: This plugin facilitates terminating non-responsive monitoring processes.
- **Sampler plugin**: Provides detailed CPU and memory sampling options.
- **Startup Profiler**: This plugin facilitates the use of instrumented profiling from startups. This is useful for applications with a short runtime.
- **Thread Inspector**: Provides advanced thread analysis.
- **Tracer**: This is a framework for detailed monitoring, using probes.
- **VisualGC**: Provides deep analysis and visualization for the garbage collector.

There are additional plugins and extensions for the VisualVM Profiler. In addition, developers can create their own custom plugins if needed.

# A comparative analysis of profiling tools

One of the great things about using profiling tools is that we have several options available to us. This section reviews the top six profiling tools featured in this chapter, using the following categories:

- Performance overhead
- Accuracy
- Ease of use
- Integration
- Cost

By following each category analysis, we will score each of these profiling tools to see how they stack up against each other.

## Performance overhead

When comparing profiling tools, it is important to consider their performance overhead. This factor can help you select the right tool for your use case. Here is a review of the performance overhead for our six selected profiling tools:

- **IntelliJ IDEA Profiler**: This tool has moderate overhead and is suitable for development and testing; however, it is not ideal for production environments, especially with high-load applications.

- **JMC**: Minimal overhead is one of JMC's strongest features. This is especially evident when using JFR. JMC is designed for production use and typically has negligible impact on performance.

- **JProfiler**: This tool has moderate overhead but a high level of detailed profiling analysis, giving developers a tough decision regarding the balance between detailed insights and increased overhead. Profiling with this tool is appropriate for development and testing environments and controlled production instances.

- **JVisualVM**: This profiling tool's performance overhead varies between low and moderate, based on the depth of profiling desired. It is suitable for development and production environments.

- **NetBeans Profiler**: This tool has moderate overhead and is only suitable for non-production environments.

- **YourKit Java Profiler**: This tool has a high level of configurability to help manage overhead. It is appropriate for both production and non-production environments.

## Accuracy

Not all profiling tools are 100% accurate, and it is important to have a firm understanding of your tool's accuracy level. Here is a review of the accuracy of your featured profiling tools:

- **IntelliJ IDEA Profiler**: This tool's greatest accuracy is with CPU and memory profiling. Accuracy takes a dip when profiling threads, and it can increase performance overhead.

- **JMC**: This is a high-accuracy profiling tool that delivers precise results with minimal performance impact, even in production environments.

- **JProfiler**: This profiling tool provides highly accurate results along with detailed visualizations.

- **JVisualVM**: While this profiling tool provides accurate sampling and data, the need for high precision increases performance overhead.

- **NetBeans Profiler**: This tool provides accurate CPU and memory profiling and real-time thread analysis.

- **YourKit Java Profiler**: This profiling tool renders highly accurate results in CPU, memory, and thread analysis.

## Ease of use

Ease of use is another important factor that we need to consider when selecting a profiling tool. If a tool is difficult to use and takes considerable time, it might not be the right tool for you. Here is a review of the ease of use for our featured tools:

- **IntelliJ IDEA Profiler**: This tool is seamlessly integrated into the IntelliJ IDEA IDE, making it intuitive and easy to use.

- **JMC**: This is an advanced tool that can require a concerted effort to learn.

- **JProfiler**: This tool is renowned for its ease of use. It is also thoroughly documented.

- **JVisualVM**: This profiling tool is beginner-friendly and has a simplistic user interface.

- **NetBeans Profiler**: This tool is integrated into the NetBeans IDE, making it easy to use.

- **YourKit Java Profiler**: This profiling tool has an intuitive user interface and is user-friendly.

## Integration

Developers tend to favor profiling tools that have deep integrations. Here is a review of that characteristic for each of our six selected profiling tools:

- **IntelliJ IDEA Profiler**: This tool is deeply integrated into the IntelliJ IDEA IDE and supports custom workflows.

- **JMC**: JMC is bundled with the JDK, making integration with JVM-based applications seamless.

- **JProfiler**: This tool easily integrates with the most popular build tools and IDEs.

- **JVisualVM**: This tool is integrated with the IDE and is easy to use.

- **NetBeans Profiler**: This tool is fully integrated into the NetBeans IDE.

- **YourKit Java Profiler**: This profiling tool supports integration with the major IDEs and CI/CD.

## Cost and licensing considerations

The cost of profiling tools can be a significant factor. Independent developers tend to favor free tools, while large teams might opt to pay for more robust tools. Here is the cost factor for our six tools:

- **IntelliJ IDEA Profiler**: This profiler is included with the IntelliJ IDEA Ultimate edition, which requires a paid subscription.

- **JMC**: This tool is free for development and testing but might require payment for use in production.

- **JProfiler**: This is a commercial tool that requires payment for use.

- **JVisualVM**: This tool is free and bundled with the JDK.

- **NetBeans Profiler**: This is a free tool that is integrated into the NetBeans IDE.

- **YourKit Java Profiler**: This is a commercial tool that requires payment for use.

## The best use cases for each tool

It is likely that no single profiling tool will fit your needs every time. Each of our tools has a specific use case they are most suited for:

- **IntelliJ IDEA Profiler**: This tool is best suited for IntelliJ IDEA users who want to use an integrated profiling tool during development

- **JMC**: This tool is ideal for production environments because of its low overhead and its ability to provide detailed analysis, especially when using JFR

- **JProfiler**: This profiling tool is ideal for developers who want an easy-to-use, powerful tool that has advanced analysis capabilities

- **JVisualVM**: This tool is best suited for developers who want a free profiling option for CPU and memory analysis during the software development and testing phases

- **NetBeans Profiler**: This tool is ideal for NetBeans IDE users who want a reliable integrated profiling tool for CPU, memory, and thread analysis

- **YourKit Java Profiler**: This profiling tool is best suited for users who want in-depth profiling during all phases of the software development life cycle, including production

## Scoring our profiling tools

The following table rates the six featured profiling tools based on performance overhead, analysis accuracy, ease of use, integration, and cost. The scores range from 1 to 5, with 1 being the lowest and 5 being the highest possible score.

| Tool | Overhead | Accuracy | Ease of Use | Integration | Cost | Overall Score |
|---|---|---|---|---|---|---|
| IntelliJ IDEA Profiler | 3 | 4 | 5 | 5 | 2 | 19 |
| Java Mission Control (JMC) | 5 | 5 | 4 | 5 | 3 | 22 |
| JProfiler | 4 | 5 | 5 | 5 | 2 | 21 |
| JVisualVM | 3 | 4 | 4 | 4 | 5 | 20 |
| NetBeans Profiler | 3 | 4 | 4 | 4 | 5 | 20 |
| YourKit Java Profiler | 4 | 5 | 5 | 5 | 2 | 21 |

Table 14.1: The profiling tool scoring matrix

As you can see, no single tool received a top score in every category. Based on this scoring, JMC received the highest score of 22, followed by JProfiler and YourKit Java Profiler at 21 points each. JVisualVM and NetBeans Profiler scored 20, and the IntelliJ IDEA Profiler scored the lowest, with 19 points.

By understanding the strengths, limitations, and trade-offs of each profiling tool, you should be able to choose the most appropriate tool for your needs.

# Practical profiling strategies

The benefits of profiling our Java applications to help identify and resolve performance issues have been established. We will next explore effective profiling strategies so that you can optimize your Java applications efficiently. This section covers strategies to identify performance bottlenecks and differentiate profiling approaches between development and production environments. We will also look at how to implement continuous profiling for long-term performance management.

## Identifying performance bottlenecks

The primary goal of profiling is to identify performance bottlenecks. We can adopt several strategies to achieve this goal, including the following:

- Begin by monitoring CPU usage, memory consumption, and response times. These high-level metrics can help you identify areas that require deeper analysis.

- Consider using sampling profilers that can quickly identify methods that consume the most CPU time. Tools such as JVisualVM and IntelliJ IDEA Profiler provide this type of functionality. The goal is to use sampling to pinpoint issues without a significant impact on performance.

- Once you have identified an area for deeper analysis, use instrumentation profiling, such as what is provided by JProfiler and YourKit Java Profiler. These tools can help you examine specific code paths and methods.

- Be sure to analyze your application's thread activity. This is critical for applications that employ concurrent processing. Tools such as JMC and NetBeans Profiler have extensive thread analysis functionality. They can be used to detect thread contention, deadlocks, and even inefficient synchronization.

- Use memory profilers to analyze object allocations, identify objects that are not collected by Java's garbage collector, and capture heap dumps. Tools such as JProfiler and YourKit Java Profiler have this capability and help identify potential memory leaks.

- Finally, if your application makes extensive use of database interactions or input/output operations, use profilers that can provide insight into JDBC calls and input/output. The aim is to identify inefficient queries and input/output bottlenecks.

## Profiling in development versus production

How and what we profile should be informed by our current environment. Profiling in a development environment should be approached differently than profiling a running application in production.

The following table provides a comparative analysis of profiling in development and production environments, using 10 key aspects of profiling.

| Aspect | Development profiling | Production profiling |
| --- | --- | --- |
| Primary focus | Identify and resolve performance issues | Monitor and resolve runtime performance issues |
| Access requirements | Direct access | Remote access |
| Frequency | Frequent profiling | Selective profiling based on observations |
| Performance impact | High overhead tolerated | Minimal overhead required |
| Profiling type | Detailed instrumentation | Sampling |
| Tools | IntelliJ IDEA Profiler, JProfiler, and YourKit Java Profiler | JMC and JVisualVM |

| Aspect | Development profiling | Production profiling |
|---|---|---|
| Data granularity | High | Low |
| Load simulation | Simulate realistic loads | Real-time user load |
| Automation | Integrated with CI/CD | Continuous monitoring |

Table 14.2: Profiling comparative analysis

Understanding the differences in profiling based on your application's environment can help you conduct efficient profiling, subsequently allowing you to enhance your Java application's performance.

## Continuous profiling

We should profile throughout the development process and then continuously once our application is in production. Implementing continuous profiling includes the following aspects:

- Establishing performance baselines so that you can compare future profiling results to them
- Integrating profiling with your CI/CD pipelines
- Ensuring you are monitoring key performance metrics including CPU use, memory use, and response times
- Storing your polling data so that you can conduct historical analysis
- In addition to routine monitoring, conducting full profiling audits at regular intervals
- Ensuring to communicate profiling results with the development team so that they are informed and can enhance the performance of your current system, helping to make future applications more performant

These practical profiling strategies can help identify and resolve performance bottlenecks. It is important to establish a formal profiling approach that can be meticulously followed.

## Case studies

This section details three case studies to illustrate the practical implementation of profiling tools in real-world scenarios.

### Case study 1 – profiling a large enterprise application

**Scenario**: A large global financial services company developed, tested, and deployed an enterprise application to handle their transactions and reporting. Recently, the application started experiencing performance degradation. This was especially evident during peak user transaction times.

**Profiling tools**: The development team selected JProfiler for its capability to provide detailed profiling and because it integrated with their development environment.

**Process implemented**: The company took a three-step approach to profiling:

1.  They first conducted an initial analysis, looking at CPU use, memory use, and thread analysis.
2.  Next, they used the profiler data from *step 1* to identify bottlenecks.
3.  The third step was to optimize their enterprise application. This optimization included rewriting inefficient algorithms, implementing more efficient data structures, optimizing object creation and disposal, and reducing thread contention by implementing a refined locking mechanism.

**Outcome**: Transaction times significantly improved, and memory consumption was now stable. Moreover, the improved application was now able to handle peak load times without network latency issues or impact on the user experience.

## Case study 2 – performance tuning in a microservices architecture

**Scenario**: A technology company developed a large-scale e-commerce application using a microservices architecture. There are a lot of microservices with the primary ones handling user authentication, product inventories, product catalogs, payment transactions, and order processing. Staff and users have reported intermittent latency issues and occasional timeouts.

**Profiling tools**: The tech company selected JMC for profiling, based on its low overhead and ability to monitor applications in production using JFR.

**Process implemented**: The company decided to enable JFR recordings on all microservices. Their plan was to collect detailed performance data without a significant impact on performance. Next, they used JMC to analyze the JFR data to identify resource usage patterns and potential hotspots.

Analysis revealed their product catalog service had sporadic CPU usage spikes that impacted overall response times. Their thread profiling analysis shows that the order processing service caused timeouts, based on thread contention. They also reviewed network profiling data, which showed the payment transaction service's database interactions had high latency.

With profiling and analysis complete, the development team optimized their e-commerce application. Specifically, they optimized the product catalog service by implementing a caching mechanism. The order processing service was rewritten with an optimized thread management schema, using thread pools. Lastly, the payment transaction service's database queries were optimized, and connection pooling was implemented.

**Outcome**: The technology company's e-commerce system showed improved performance following the optimizations. Latency was reduced and it was significantly more stable, with few timeouts. The staff reported improved satisfaction and user complaints waned.

## Case study 3 – profiling and optimizing a high-throughput system

**Scenario**: A communications company uses a system for high throughput and real-time data processing and analytics. Their system has significant performance degradation during high load times. This leads to delays in data processing and analysis operations.

**Profiling tools**: The company selected the YourKit Java Profiler tool because it has comprehensive features, and it is ideal for high-throughput systems.

**Process implemented**: The communication company's development team adopted a three-step approach – conduct initial profiling, identify bottlenecks, and optimize.

The initial profiling included a look at CPU and memory use. They focused on data processing and analytic components of their system. They also conducted a detailed thread analysis to help identify bottlenecks evident in thread processing.

Their second step was to identify bottlenecks. Their CPU profiling reviewed that some of their data processing modules contained complex computations, which have a significant draw on CPU use. Their memory profiling results indicated that their analytics module had high memory usage. That same module showed frequent garbage collection events that negatively impacted the system's overall performance. They also looked at thread analysis results, which showed thread contention during peak data loads, further reducing performance.

The company's last step was to optimize its system. They optimized their data processing algorithms to improve the efficiency of their computations and enhanced parallel processing to take advantage of multi-core processors. Developers also optimized their analytics module by reducing object creation, improving data structures, and minimizing garbage collection. The team also optimized how their application managed threads. They introduced thread pools to reduce thread contention.

**Outcome**: The communication company's high-throughput system realized significant performance improvements following the system optimizations. Their data processing times were reduced by 55% and latency during analytics processing was minimized. Their system can now handle high data loads efficiently and meet real-time data processing requirements.

The three case studies demonstrate a practical application of profiling strategies and tools in various contexts. They all used a similar three-step approach (profile, identify, optimize) and had successful outcomes.

## Future trends in Java profiling

The software development landscape is ever-shifting, and profiling tools and techniques are equally dynamic. As technological innovations emerge, so do tools to help developers address and manage them. This final section explores future trends in Java profiling, with a focus on profiling tools and integration with AI and ML. The section ends with a list of emerging standards and best practices.

# Advances in profiling tools

Current profiling tools continue to be improved, with new versions frequently being released. It is also common for new profiling tools to be introduced periodically. The key advancements with profiling tools are in the following five areas:

- **Improved user interfaces**: We should see more intuitive interfaces and better visualizations.

- **Cloud-native profiling**: Cloud computing and distributed computing are the norm, and profiling tools are apt to cater to these environments to a greater extent.

- **Enhanced real-time profiling**: Profiling tools will enhance their real-time processing capabilities and continually decrease their impact on performance.

- **Low-overhead instrumentation**: Profiling high-throughput systems can result in system latency. Future advances will decrease the impact these systems have on performance.

- **Unified monitoring and profiling**: The convergence of monitoring and profiling tools can be advantageous for development teams that want diagnostics that combine real-time monitoring and deep profiling data.

# Integration with AI and machine learning

AI and ML technologies can be leveraged by profiling tools. These technologies can help performance-tuning efforts. Here are key aspects of the use of AI and ML in performance tuning:

- Adaptive profiling

- Automated anomaly detection

- Intelligent optimization recommendations

- Predictive performance modeling

- Root cause analysis

# Emerging standards and best practices

The following recommendations are intended to help guide your effective use of profiling tools and techniques as they exist today and as they evolve:

- Adopt a continuous profiling strategy.

- Be mindful of security and privacy considerations when profiling tools collect data.

- Communicate and collaborate with your team. This includes their involvement in your profiling strategy, a review of the results, and optimization planning.

- Include your profiling efforts in a holistic performance management plan.

- Standardize your profiling APIs so that you can introduce interoperability and integration simplification. These standards can help ensure that your profiling data is consistent and that you can effectively compare the standards across different tools and environments.

It is important to stay abreast of changes to the profiling tools you use and to be cognizant of new tools and techniques as they are introduced. This awareness, coupled with adherence to best practices, can help ensure that you fully leverage profiling tools and techniques, resulting in high-performing Java applications.

## Summary

This chapter took a thorough look at profiling tools and their role in Java application performance. We explored various profilers, including those bundled in the JDK and embedded in IDEs, and third-party tools such as JProfiler and YourKit Java Profiler. Coverage included practical profiling strategies to help identify performance bottlenecks, distinct profiling approaches required for development and performance environments, and the importance of continuous profiling to support long-term performance management. We introduced three real-world case studies to illustrate profiling tool applications. Finally, we examined future trends and best practices regarding Java profiling tools.

In the next chapter, we will examine how to optimize our databases and queries to enhance the performance of our Java applications. We will review database design considerations, purposeful SQL query generation, and several strategies, including normalization, indexing, connection pooling, caching, JDBC and ORM optimizations, transaction management, and testing.

# Part 5:
# Advanced Topics

The final part of the book addresses advanced topics that push the boundaries of Java performance optimization. It begins with strategies to optimize databases and SQL queries, followed by techniques for effective code monitoring and maintenance. You will learn about unit and performance testing to ensure that your applications meet performance standards. This part concludes with an exploration of leveraging **artificial intelligence (AI)** to enhance the performance of Java applications, providing a forward-looking perspective on future trends and technologies.

This part has the following chapters:

- *Chapter 15, Optimizing Your Databases and SQL Queries*
- *Chapter 16, Code Monitoring and Maintenance*
- *Chapter 17, Unit and Performance Testing*
- *Chapter 18, Leveraging Artificial Intelligence (AI) for High Performance Java Applications*

# 15

# Optimizing Your Database and SQL Queries

Databases are a key component of large software systems. These systems constantly retrieve and update data using database connections and queries. The scale of data in modern systems is extremely large, resulting in more data to query, update, and display. This increased scale can result in negative performance issues for our Java application, underscoring the significance of ensuring our databases and queries are optimized.

This chapter examines critical database design concepts, including database schemas, indexing strategies, and data partitioning techniques. Database queries are also examined, with a specific focus on **Structured Query Language (SQL)** queries. Our coverage of query optimizations includes best practices for writing efficient queries, query execution planning, and advanced SQL techniques.

The chapter also covers advanced SQL techniques such as database configuration, performance monitoring, and database maintenance. The chapter ends with several real-world case studies that help demonstrate how to identify and resolve database-related performance issues in existing systems.

By the end of this chapter, you should have a foundational understanding of strategies to optimize your databases and database queries. Armed with this understanding, and leveraging your experience gained from hands-on exercises, you should be able to improve the performance of your Java applications that incorporate databases and database queries.

This chapter covers the following main topics:

- Database design
- SQL query optimizations
- Additional strategies
- Case studies

# Database design

For new systems, we have the luxury of designing our databases with performance in mind. Our database's design can have a significant impact on the efficiency of our SQL queries. In this section, we will examine key principles of database design, including schema, indexing, partitioning, and sharding.

## Schema principles

> **Database schema**
>
> A database schema is the design of the database, serving as a blueprint to create it.

Before we create our database, we should create a **schema** to document how our data will be organized and to indicate how it is interrelated. Our goal is to design a schema that makes querying the database more efficient.

An early decision to make is whether our database will be **normalized** or **denormalized**. A denormalized database involves reducing the number of tables to decrease the complexity of queries. Conversely, normalization involves creating separate tables to eliminate duplicative data. Let's look at an example.

The following table shows duplicative data for both the author and publisher fields.

| BookID | Author | Title | Publisher | Price ($) |
|--------|--------|-------|-----------|-----------|
| 1 | N. Anderson | *Introduction to Zion* | Packt | 65.99 |
| 2 | N. Anderson | *Illustrated History of Zion* | Packt | 123.99 |
| 3 | W. Rabbit | *Astro Mechanics* | Packt | 89.99 |
| 4 | W. Rabbit | *Gyro Machinery* | Forest Press | 79.99 |

Table 15.1 – A Denormalized table

As you can see in the preceding table, there are two entries for two different authors, and one publisher is listed more than once. This is an unnormalized table. To normalize the table, we will create three tables, one each for the books, authors, and publishers.

| BookID | Title | AuthorID | PublisherID | Price ($) |
|--------|-------|----------|-------------|-----------|
| 1 | *Introduction to Zion* | 1 | 1 | 65.99 |
| 2 | *Illustrated History of Zion* | 1 | 1 | 123.99 |

| BookID | Title | AuthorID | PublisherID | Price ($) |
|---|---|---|---|---|
| 3 | *Astro Mechanics* | 2 | 1 | 89.99 |
| 4 | *Gyro Machinery* | 2 | 2 | 79.99 |

Table 15.2 – The Books table

The `Books` table references the `AuthorID` and `PublisherID` fields. Those are established in the following tables. Here is the `Authors` table:

| AuthorID | Author |
|---|---|
| 1 | N. Anderson |
| 2 | W. Rabbit |

Table 15.3 – The Authors table

Our final table is for the **Publishers**:

| PublisherID | Publisher |
|---|---|
| 1 | Packt |
| 2 | Forest Press |

Table 15.4 – The Publishers table

The decision to implement a normalized or denormalized database involves considering the complexity of queries, the size of your database, and the read-write load on your database.

Another important database design consideration is the data type for each of your columns. For example, it is appropriate to use an **integer** (**INT**) for our ID fields instead of a less efficient implementation. Here is how we would create our `Authors` table with appropriate data types using SQL:

```
CREATE TABLE Authors {
  Author_ID INT,
  Author VARCHAR(80)
};
```

After we have designed our database tables and decided on data types, we need to implement indexing. Let's look at that in the next section.

## Indexing

We index our databases so that data can be found quickly. Our indexing strategy has a direct impact on our query performance, so due diligence is required. There are two types of indexing. The first type is **balanced tree (B-tree)**, which is what is implemented in most databases. This type of index keeps data sorted and permits sequential access, insertions, and deletions.

The second type of database indexing is **hash indexes**. This type of indexing is ideal when equality comparisons are needed but is not adequate for range queries.

Index creation is simple and demonstrated by the following SQL statement:

```
CREATE INDEX idx_authors_authorid ON Authors (AuthorID);
```

If you have an existing database and frequently use WHERE, JOIN, ORDER BY, or GROUP BY, you can likely benefit from indexing. Let's look at an example.

We could use the following SQL statement when we do not have an index:

```
SELECT * FROM Authors WHERE Author = 'N. Anderson';
```

The following SQL statement searches for the same author but uses an index:

```
CREATE INDEX idx_authors_authorid ON Authors (AuthorID);
SELECT * FROM Authors WHERE Author = 'N. Anderson';
```

Using the second example will provide results a bit faster than the non-indexed approach. Note that indexes do take up additional storage space and add additional processing overhead when using the INSERT, DELETE, and UPDATE operations.

In the next section, we will examine partitioning and sharding as approaches to improve the efficiency of our database and database queries.

## Partitioning and sharding

Partitioning and sharding are strategies used to improve the performance of large databases and their queries by dividing large datasets into smaller components.

### Partitioning

There are two types of partitioning, **horizontal partitioning** and **vertical partitioning**. Horizontal partitioning is accomplished by splitting tables into rows, and each horizontal partition contains a subset of those rows. A typical use case for this is creating a partition based on date ranges. The following example creates three tables, each with a specific year's worth of order information:

```
CREATE TABLE Book_Orders_2021 (
    CHECK (OrderDate >= '2021-01-01' AND OrderDate < '2022-01-01')
```

```
) INHERITS (Book_Orders);

CREATE TABLE Book_Orders_2022 (
    CHECK (OrderDate >= '2022-01-01' AND OrderDate < '2023-01-01')
) INHERITS (Book_Orders);

CREATE TABLE Book_Orders_2023 (
    CHECK (OrderDate >= '2023-01-01' AND OrderDate < '2024-01-01')
) INHERITS (Book_Orders);
```

Vertical partitioning splits the database table into columns, each partition containing a subset of columns. To demonstrate vertical partitioning, let's look at a Books table that has not been partitioned:

```
CREATE TABLE Books (
    BookID INT PRIMARY KEY,
    Title VARCHAR(100),
    AuthorID INT,
    Pages INT,
    Genre VARCHAR(50),
    PublisherID INT,
    PublishedDate DATE,
    OriginalPrice DECIMAL(10, 2),
    DiscountPrice DECIMAL(10, 2)
);
```

Now, using vertical partitioning, let's create two tables, each with a subset of columns:

```
CREATE TABLE Books_CatalogData (
    BookID INT PRIMARY KEY,
    Title VARCHAR(100),
    AuthorID INT,
    Pages INT,
    Genre VARCHAR(50),
    PublisherID INT,
    PublishedDate DATE,
);
CREATE TABLE Books_SalesData (
    BookID INT PRIMARY KEY,
    OriginalPrice DECIMAL(10, 2),
    DiscountPrice DECIMAL(10, 2),
    FOREIGN KEY (BookID) REFERENCES Books_CatalogData(BookID)
);
```

By splitting our table into two partitions, we can search more efficiently, since we do not need to process any unrelated data (i.e., when searching the catalog data, we are not concerned with sales data).

Let's now look at another strategy to increase the efficiency of our databases, called sharding.

### Sharding

**Sharding** is the process of distributing our data over multiple servers, moving the data closer to users. This strategy has two primary benefits – moving the data to servers closer to users reduces network latency and the load on individual servers. A common use case is to shard based on geographic region. Here is an example of how we can accomplish that:

```
CREATE TABLE Users_US (
    UserID INT PRIMARY KEY,
    UserName VARCHAR(100),
    Region CHAR(2) DEFAULT 'US'
);

CREATE TABLE Users_EU (
    UserID INT PRIMARY KEY,
    UserName VARCHAR(100),
    Region CHAR(2) DEFAULT 'EU'
);
```

The preceding example creates two tables. The next step would be to store each table on different servers.

A purposeful approach to partitioning and sharding can result in a performance-ready database design. It can make your SQL queries more efficient, thereby improving the overall performance of your Java application.

# Query optimizations

Now that we have a basic understanding of how to design our databases with performance in mind, we are ready to look at best practices for writing efficient queries. We will also look at query execution plans and some advanced SQL techniques. To make our example SQL statements relatable, we will use a book inventory and order processing database throughout this section.

## Query execution

Understanding how queries are handled by our database is key to being able to optimize them.

> **A query execution plan**
> A query execution plan provides details on how a database engine executes queries

A query execution plan includes details on database query operations, such as joins and sorts. Let's look at a simple query that gives us a specific book's total sales:

```
FROM Orders o
JOIN Books b ON o.BookID = b.BookID
WHERE b.Title = 'High Performance with Java';
```

Now, let's add an EXPLAIN command to the same query to reveal the steps the database engine follows to execute our query:

```
EXPLAIN SELECT SUM(o.Qty * o.Price) AS TotalSales
FROM Orders o
JOIN Books b ON o.BookID = b.BookID
WHERE b.Title = 'High Performance with Java';
```

By viewing the query execution plan, we can identify potential bottlenecks, providing us with an opportunity to further optimize our database and queries.

Next, let's look at some best practices for writing efficient queries.

## Best practices

Our goals when writing SQL queries are to minimize resource use and reduce execution time. To achieve these goals, we should follow best practices, including the ones detailed as follows for the SELECT statement, JOIN operations, **subqueries**, and **common table expressions (CTEs)**.

### SELECT statement

There are three best practices involved with using the SELECT statement. First, we should avoid using SELECT * and only specify the columns we need. For example, instead of using SELECT * FROM Books;, use SELECT Title, AuthorID, Genre FROM Books;.

Another best practice is to use the WHERE clause to narrow down our results to the maximum extent possible. Here is an example:

```
SELECT Title, AuthorID, Genre FROM Books WHERE Genre = 'Non-Fiction';
```

A third best practice for using the SELECT statement is to limit the number of rows returned by our query. We can use the LIMIT clause, as shown here:

```
SELECT Title, AuthorID, Genre FROM Books WHERE Genre = 'Non-Fiction'
LIMIT 10;
```

The three best practices for working with the SELECT statement are key to improving the efficiency of our queries. Next, let's look at best practices for using JOIN operations.

## JOIN operations

There are two best practices for using JOIN operations. First, we should ensure that all columns used in JOIN conditions are indexed. This will improve the efficiency of these operations.

Another best practice is to use the appropriate JOIN type, as indicated in the following table.

| Type | Purpose |
|------|---------|
| INNER JOIN | Used to match rows |
| LEFT JOIN | To include all rows from the left table |
| RIGHT JOIN | To include all rows from the right table |

Table 15.5: JOIN type and its purpose

Next, let's look at the concept of subqueries and their related best practices.

## Subqueries

As the title suggests, **subqueries** are used to break a complex query into multiple, simpler queries. Here is an example:

```
SELECT b.Title, b.Genre
FROM Books b
WHERE b.BookID IN (SELECT o.BookID FROM Orders o WHERE o.Qty > 100);
```

Next, let's look at CTEs.

## CTEs

CTEs can be used to make complex queries more readable. This increases their reusability and eases their maintainability. Here is an example:

```
WITH HighSales AS (
    SELECT BookID, SUM(Qty) AS TotalQty
    FROM Orders
    GROUP BY BookID
    HAVING SUM(Qty) > 100
)
SELECT b.Title, b.Genre
FROM Books b
JOIN HighSales hs ON b.BookID = hs.BookID;
```

Now that we have reviewed several best practices for writing queries, let's look at some advanced SQL techniques.

## Advanced SQL techniques

This section demonstrates three advanced SQL techniques – window functions, recursive queries, and temporary tables and views.

### Window functions

A **window function** is used to calculate across a set of rows related to a current row. Here is an example:

```
SELECT BookID, Title, Genre,
       SUM(Quantity) OVER (PARTITION BY Genre) AS TotalSalesByGenre
FROM Books b
JOIN Orders o ON b.BookID = o.BookID;
```

### Recursive queries

Recursive queries are complicated and can be useful when you have hierarchical data, such as book categories and subcategories. Here is an example:

```
WITH RECURSIVE CategoryHierarchy AS (
    SELECT CategoryID, CategoryName, ParentCategoryID
    FROM Categories
    WHERE ParentCategoryID IS NULL
    UNION ALL
    SELECT c.CategoryID, c.CategoryName, c.ParentCategoryID
    FROM Categories c
    JOIN CategoryHierarchy ch ON c.ParentCategoryID = ch.CategoryID
)
SELECT * FROM CategoryHierarchy;
```

### Temporary tables and views

Another advanced technique is to use temporary tables and views to achieve better performance and help manage complex queries. Here is an example of a temporary table:

```
CREATE TEMPORARY TABLE TempHighSales AS
SELECT BookID, SUM(Qty) AS TotalQty
FROM Orders
GROUP BY BookID
HAVING SUM(Qty) > 100;

SELECT b.Title, b.Genre
FROM Books b
JOIN TempHighSales ths ON b.BookID = ths.BookID;
```

The following SQL statement is an example of a temporary view:

```
CREATE VIEW HighSalesView AS
SELECT BookID, SUM(Qty) AS TotalQty
FROM Orders
GROUP BY BookID
HAVING SUM(Qty) > 100;


SELECT b.Title, b.Genre
FROM Books b
JOIN HighSalesView hsv ON b.BookID = hsv.BookID;
```

Experimenting with the advanced techniques presented in this section can improve your ability to write efficient queries, contributing to the overall performance of your Java application.

# Additional strategies

So far, this chapter has covered designing a database schema for efficiency and how to write efficient SQL queries. There are several additional strategies we can employ, including fine-tuning, monitoring, and maintenance. Each of these strategies is explored in this section and uses the same book inventory and ordering example from the previous section.

## Fine-tuning

We can fine-tune our database server's configuration parameters to ensure that our queries make efficient use of resources. This fine-tuning can be categorized as follows:

- **Database server parameters**
- **Memory allocation**: It is important to ensure we allocate sufficient memory for buffering and caching. For example, we can adjust the innodb_buffer_pool_size parameter in MySQL with the SET shared_buffers = '3GB'; SQL statement.
- **Connection pooling**: As detailed in *Chapter 10, Connection Pooling*, we can pool our database connections to reduce overhead and improve overall application performance.
- **Query caching**: Query caching can be used to store the results of commonly executed queries. Using an SQL statement such as SET query_cache_size = 256MB'; will enable query caching.
- **Memory management**
- **Database caching**: We can cache databases to speed up read operations for frequently accessed data. Tools such as **Redis** can be used to aid in this technique.

Next, let's explore how we can monitor and profile our database's performance.

## Database performance monitoring

Once our database is up and running and all our queries are established, we are ready to monitor our database's performance. Monitoring can help us identify potential bottlenecks and allow us to make refinements to improve overall performance.

A proven approach to identify bottlenecks is to enable **slow query logging**. This can help us identify which queries take longer than we desire to execute. Here is how this can be enabled:

```
SET long_query_time = 1;
SET slow_query_log = 'ON';
```

The use of query profiling tools can help us analyze and optimize slow queries. There are various tools available, depending on your database type and service.

Monitoring and profiling can help identify opportunities for refinement. In the next section, we will explore database maintenance.

## Database maintenance

Databases are dynamic and need to be maintained with regularly scheduled maintenance. This is a proactive, vice reactive approach to maintaining your databases. Here are some tips:

- Regularly run VACUUM to reclaim storage.
- After each run of VACUUM, run ANALYZE so that the query planner has updated statistics.
- Use REINDEX to periodically reindex your database. This will improve query performance.
- Archive old data that is no longer needed. You can partition this data into a historical database.
- Purge data that is not needed. This will free up storage and should improve query performance.

The strategies presented in this section can help you further enhance your query performance, the database performance, and the overall performance of your Java application. These strategies are especially important for large databases and those with high transaction rates.

In the next section, we will review several real-world case studies to help you contextualize the concepts presented in this chapter.

# Case studies

This section presents three real-world case studies using the book inventory and order processing database featured throughout this chapter. A review of the case studies will demonstrate how the strategies and techniques presented in this chapter can be used to solve common database performance problems.

Each case study is presented in the following format:

- The scenario
- The initial SQL query
- The problem
- The optimization steps
- The result

## Case study 1

**Scenario**: Every time the bookstore's administrator runs the sales report, it takes several minutes – much longer than it should. The report simply summarizes total sales by title. The database schema is the same as the one presented earlier in this chapter.

**Initial SQL query**:

```
SELECT b.Title, SUM(o.Qty * o.Price) AS TotalSales
FROM Orders o
JOIN Books b ON o.BookID = b.BookID
GROUP BY b.Title;
```

**Problem**: The query performs a full table scan of both the `Books` and `Orders` tables. This results in slow performance.

**Optimization steps**:

1. The database administrator added indexes to the `BookID` columns in both the `Books` and `Orders` tables:

   ```
   CREATE INDEX idx_books_bookid ON Books (BookID);
   CREATE INDEX idx_orders_bookid ON Orders (BookID);
   ```

2. The query was refined to only include columns that were needed to retrieve the desired data:

   ```
   EXPLAIN ANALYZE
   SELECT b.Title, SUM(o.Qty * o.Price) AS TotalSales
   FROM Orders o
   JOIN Books b ON o.BookID = b.BookID
   GROUP BY b.Title;
   ```

**Result**: The execution plan, revealed by using the `EXPLAIN ANALYZE` command, showed a significant query time reduction. The sales report now runs in less than one minute.

## Case study 2

**Scenario**: The database has grown exponentially, and the `Orders` table now contains millions of records. Running queries for specific years is extremely slow.

**Initial SQL query**:

```
SELECT * FROM Orders WHERE OrderDate BETWEEN '2023-01-01' AND '2023-
12-31';
```

**Problem**: The query performs a full table scan, which results in slow performance.

**Optimization steps**:

1.  The database administrator performed horizontal partitioning, creating tables for each year:

    ```
    CREATE TABLE Orders_2023 (
        CHECK (OrderDate >= '2023-01-01' AND OrderDate < '2024-01-
        01')
    ) INHERITS (Orders);

    CREATE TABLE Orders_2024 (
        CHECK (OrderDate >= '2024-01-01' AND OrderDate < '2025-01-
        01')
    ) INHERITS (Orders);
    ```

2.  After partitioning the data, the administrator updated the queries to target the specific partitions:

    ```
    SELECT * FROM Orders_2023 WHERE OrderDate BETWEEN '2023-01-01'
    AND '2023-12-31';
    ```

**Result**: The query performance significantly improved, running at a fraction of the previous time.

## Case study 3

**Scenario**: The bookstore has a web application that is used to display book data on product pages. The repeated querying of the `Books` table resulted in a significant and unnecessary load on the database.

**Initial SQL query**:

```
SELECT * FROM Books WHERE BookID = 1;
```

**Problem**: Identical queries were repeatedly sent to the database, creating a high load and slow response time.

**Optimization step**: The database administrator used **Redis** to cache book details.

**Result**: The database load was significantly reduced, and the response times were drastically shorter.

## Summary

This chapter explored essential strategies and techniques to optimize databases and SQL queries. The chapter's overall aim was to introduce database-related enhancements and best practices to improve the performance of your data-driven applications. We began with the fundamentals of database design, including schema normalization, appropriate indexing, and partitioning strategies. We then explored how to write efficient SQL queries. Our coverage also included query execution plans and leveraging advanced SQL techniques, such as window functions and recursive queries. Additional strategies, including database configuration, monitoring, profiling, and regular maintenance, were also discussed. The chapter ended with real-world case studies to demonstrate the practical application of the strategies and techniques covered in the chapter. You should now be confident in implementing these best practices and ensuring your database systems can handle large datasets and complex queries with ease.

In the next chapter, we will examine the concepts of code monitoring and code maintenance, with an ever-vigilant eye on the high performance of our Java applications. Our approaches to code monitoring and maintenance will include conducting code reviews to identify potential performance issues before they become problematic. Specifically, we will look at **application performance management (APM)** tools, code reviews, log analysis, and continuous improvement.

# 16

# Code Monitoring and Maintenance

Writing efficient code is not enough to ensure that our Java applications perform at a high level after the initial launch of our systems. We must adopt a robust strategy of coding monitoring and maintenance to ensure that our systems continue to perform at desired levels even as data and use volumes increase and environments change. This chapter focuses on the critical practices, and associated tools, of code monitoring and maintenance.

Our chapter starts with an exploration of **Application Performance Management (APM)** tools that we can use to conduct real-time monitoring and obtain diagnostic data to help us keep our applications running efficiently. Our APM tool exploration will include use cases and implementation strategies.

The importance of **code reviews** is also covered with the goal of instilling a dedication to continual process improvement, specifically to continually maintain code quality. Insights into best practices and automation tools will provide you with knowledge on how you use selected tools and practices to identify potential issues in your code before they cause undesired system behavior and negatively impact the user experience.

The chapter also introduces the concept of logging and shares how effective it can be for monitoring applications. We will explore best practices, logging frameworks, and how to analyze log data. Our goal is to log the correct data and learn to use the logged data to identify optimization opportunities without introducing excessive system overhead.

This chapter covers the following main topics:

- APM tools
- Code reviews
- Logging
- Monitoring and alerting
- Maintenance strategies

# APM tools

Java developers spend more time maintaining their systems than they do developing them. This is because our systems are in production longer than it takes to write and test our code. It stands to reason that we should approach **Application Performance Management** (**APM**) seriously and appreciate the crucial role it has in our ability to ensure our systems continue to perform optimally and maximize the user experience.

## APM tool overview

APM tools are designed to help us monitor our systems to manage their performance. These tools provide real-time insights into how our applications are performing. They can help us identify potential bottlenecks and possible optimization opportunities such as with resource use. The metrics generated by APM tools can be measured against baselines and give us a true picture of our system's overall health.

The primary objectives of APM tools include the following:

- Ensuring application reliability
- Ensuring application scalability
- Identifying performance issues
- Providing actionable insights
- Real-time application monitoring
- User interaction tracking

Let's next look at the key features of APM tools.

## APM tool key features

AMP tools range from basic to robust and can have a varied array of features. The top 10 key features of AMP tools are as follows, in no specific order:

- Provide configurable alerts based on performance thresholds
- Monitor system resource usage
- Continuously monitor application performance such as response times
- Identify performance bottlenecks by tracing individual transactions
- Identify and log exceptions and errors with sufficient detail to support troubleshooting
- Track and measure user interactions such as transaction durations and load times
- Conduct in-depth analysis

- Provide numeric and visual reporting of analytic data

- Integrate with other tools such as CI/CD pipelines

- Automate performance management processes

Now that we have an appreciation for the importance of APM tools, their objectives, and their key features, let's review five common APM tools used for Java application monitoring.

## Popular APM tools

There is a plethora of AMP tools available to Java application developers. The table that follows lists five common tools and provides a brief description and URL for each.

| Tool | Description | URL |
|---|---|---|
| **AppDynamics** | Strong performance monitor with great visibility into Java applications | `https://www.appdynamics.com/product/application-performance-monitoring` |
| **Datadog** | Cloud-based monitoring and analytics tool | `https://www.dynatrace.com/monitoring/solutions/cloud-monitoring-cio-report/` |
| **Dynatrace** | Advanced tool that leverages AI for performance monitoring | `https://www.dynatrace.com/monitoring/platform/application-observability/` |
| **Elastic APM** | Open source performance monitoring tool | `https://www.elastic.co/observability/application-performance-monitoring` |
| **New Relic** | Perhaps the most commonly used tool, with extensive monitoring and analytics capabilities | `https://www.newrelic.com` |

Table 16.1: Popular APM tools for Java applications

**Note on URLs**

The URLs listed in the preceding table were valid at the time of this book's initial publication date. If you find a link that is no longer valid, you can search for the tool name along with *APM tool for Java applications* (for example, *Dynatrace APM tool for Java applications*) to find the new link.

You should take the time to experiment with each APM tool to determine which one or ones you'd like to work with. Once you have an idea of which one(s) you will use, you can review the best practices presented in the next section.

## APM tool best practices

Regardless of which APM tool you adopt, there are several best practices that can help you leverage your selected tool to support your monitoring efforts. Here are several best practices for you to consider:

- Regularly analyze your performance data. This can help you identify trends and anomalies.
- Schedule and conduct periodic reviews of your APM configurations and performance data.
- Establish clear performance monitoring goals, aligning them with your organization's objectives.
- Create your **Key Performance Indicators** (**KPIs**) and periodically review them for relevancy.
- Establish a culture of performance awareness among your teams.
- Implement a comprehensive approach to instrumentation to ensure that everything that can be monitored is.
- Integrate APM tools with your DevOps processes.
- Prioritize monitoring your most critical metrics such as error rates, response times, resource use, and so on.
- Configure your APM tool to provide you with alerts and notifications for greater efficiency.

Following the best practices presented here can help you maximize the benefits of using APM tools.

## Code reviews

Code reviews are a fundamental component of software development. As we previously suggested, code quality does not remain constant after it goes live. Environments change, new data is introduced, scaling can occur, and user behavior can change. This underscores the importance of conducting code reviews.

The purposes of code reviews are listed here:

- Ensuring consistency in adherence to standards and guidelines
- Encouraging collaboration
- Facilitating knowledge transfer among teams and increasing individual buy-in to code quality
- Increasing optimization
- Identifying defects and security vulnerabilities with quality assurance

Next, let's review some best practices when conducting code reviews.

## Best practices

Here are some self-descriptive best practices to consider when conducting code reviews:

- Automate where possible
- Encourage constructive feedback
- Establish coding standards
- Limit the size of code changes
- Prioritize critical code sections
- Set code review time limits
- Use checklists for reviewer consistency

The first best practice suggests that you automate where possible. There are automated code review tools worth your independent research and analysis. Here is a list of these to review:

- **Checkstyle**
- **Code Climate**
- **FindBugs** and its successor, **SpotBugs**
- **PMD**
- **SonarQube**

Take your time reviewing the automation tools. Once you have selected the one you like best, experiment with it before officially adopting it for live projects. Next, let's look at peer review processes.

## Peer review processes

Code review is often done by peers and it can be awkward if not approached properly. Here are some tips for conducting proper and efficient peer reviews:

- Assign reviewers who have the appropriate expertise and experience
- Rotate reviewers to mitigate complacency
- Prepare for reviews by mandating in-code commenting standards
- Use code review tools (such as GitHub pull requests) to streamline the process
- Encourage an atmosphere of mutual respect and open communication
- Follow up on all feedback

Following these tips can ease the awkwardness and improve the efficiency of peer reviews. Next, let's look at some common pitfalls regarding code reviews.

## Common pitfalls

There are several pitfalls when conducting code reviews, although they are overshadowed by the benefits we get from the process. This section examines the top five pitfalls experienced when conducting code reviews and suggests solutions.

| Pitfall | Solution |
| --- | --- |
| Delayed reviews | Set clear timelines and integrate them into your development workflow. |
| Inconsistent standards | Establish clear guidelines and use checklists. |
| Lack of constructive feedback | Focus on providing constructive feedback. Be specific and ensure that feedback is actionable. |
| Neglecting automation tools | Use automation tools to your advantage. They can catch routine issues while your developers review more complex issues. |
| Overly long reviews | Keep reviews short and frequent. |

Table 16.2: Common code review pitfalls with solutions

An understanding of these pitfalls can help ensure that your code reviews are seamlessly integrated into your workflows.

In the next section, we will review logging.

# Logging

Logging data is a fundamental component of code monitoring and maintenance. This data can reveal information about how our code performs, where it is failing, what security concerns there might be, and more. It involves recording information about how our program runs. We can use this information for audits, debugging, and monitoring.

The key aspects of logging include the following:

- **Log levels**: Logs are categorized based on their importance and the severity of what is being logged. Examples include DEBUG, ERROR, and INFO.

- **Log messages**: These are the narrative descriptions of application events.

- **Log rotation**: We rotate logs by archiving old logs and starting new ones. This prevents singular, large logs that can be difficult to manage and result in storage issues.

- **Log targets**: Targets are the storage destinations for the logs.

Now that we have a fundamental understanding of logging, let's review some best practices.

## Best practices

Our logging implementations should be focused on the relevancy of what is captured and on the efficiency of the system. With those goals in mind, here are some best practices regarding logging:

- Adopt a concise but descriptive mentality for logs. You will want them to be clear and not overly verbose.

- Centralize your logs to facilitate aggregated and comprehensive processing.

- Consider using a structured logging format such as JSON so they are easier to parse and analyze.

- Focus on log levels that are aligned with your systems' most critical processes.

- Formalize your logging practices to include formats and naming conventions.

- Protect individually identifiable or other sensitive data from being included in logs.

- Review your logs regularly.

Next, we will review useful logging frameworks.

## Logging frameworks

There are several logging frameworks that we can use for our Java applications. Developers typically select one after reviewing their options. Here is a list of some of the more popular frameworks:

- **Java Util Logging** (JUL): This is a logging framework built into Java (`java.util.logging`). While it only provides basic logging capabilities, we can use it without the need for additional libraries. It is a good framework to use if you are just getting started with logging.

- **Log4j2**: This is an advanced framework that supports various configurations, log levels, types, and destinations.

- **Logback**: This framework provides a high-performance option compatible with **Simple Logging Facade for Java (SLF4J)**.

- **SLF4J**: This framework provides an abstraction to support multiple logging frameworks such as Logback.

- **Tinylog**: As the name suggests, this is a lightweight framework with low overhead. It is ideal for small applications.

These frameworks can help simplify our implementation and management of logging. Next, let's look at key strategies for analyzing our logged data.

## Analyzing log data

One of the primary reasons for logging systems data is to provide us with the ability to analyze it for the betterment of our system's overall performance. Here are some strategies you can use for analyzing and managing your log data:

- **Aggregate**: Feed your logs into a central repository for more efficient analysis. There are several tools (such as **Logstash** and **Elasticsearch**) that can be used to help with this strategy.

- **Analyze**: Use tools (such as **Datadog** and **Graylog**) to help analyze and visualize logged data. Use statistical analysis to gain deep insights.

- **Automate**: Use automated alerts to inform you of activity based on thresholds you set.

- **Archive**: Archive logs to avoid log bloat.

Following the best practices and leveraging selected frameworks and tools can help ensure your logging efforts are purposeful and efficient. Next, let's look at how to set up monitoring and alerts.

# Monitoring and alerting

Our log data is always available when we want to review it, but more importantly, it can be used to provide us with automated alerts based on how we set things up. Effective monitoring and alerting are key operations for maintaining our Java applications and ensuring their high performance and security.

Monitoring and alerting can provide us with real-time insights into how our application is performing and promptly alert us to anything that requires immediate action.

## Monitoring system setup

The steps involved in setting up your monitoring system will depend on the frameworks and tools you select. Here is a six-step process that can be used irrespective of the frameworks and tools you select:

1. **Identify key metrics**: You need to know what you want to collect so that what is collected is useful. Your critical metrics might include CPU usage, memory use, error rates, response times, and so on. Once you have identified your key metrics, you can establish performance goals.

2. **Select monitoring tools**: Select the monitoring tools (such as **Grafana**, **New Relic**, or **Prometheus**) that are the most appropriate for your application's architecture and requirements.

3. **Integrate monitoring agents**: Integrate monitoring agents into your Java application to collect performance data. Depending on your solution, you might add specific monitoring code, leverage built-in capabilities of existing frameworks, or use APIs.

4. **Set up data collection**: Configure your monitoring system to collect and store performance data. Be mindful and ensure that the data collection operation is efficient and does not result in significant application overhead.

5. **Visualize**: Create or use dashboards provided by your monitoring tools to visually represent the collected data. You can use tools such as Grafana to build interactive dashboards that can help you quickly understand your application's performance status.

6. **Review and adjust**: Periodically review the effectiveness of your monitoring setup and make necessary adjustments. Reassess this each time you introduce a new component or service.

Once your monitoring system has been set up, you will be ready to configure it to provide informative alerts.

## Alert configuration

Configuring alerts involves setting thresholds and rules that trigger notifications when certain performance conditions are met. Follow these steps to set up an effective alert schema:

1. **Define criteria**: Identify the conditions that you want alerts for and be specific. For example, you might choose 92% CPU usage, increased response times, spikes in errors, and so on. You can base your thresholds on historical application data and performance benchmarks.

2. **Set levels**: Categorize your alerts by severity (that is, **information**, **warning**, **critical**) so your team can prioritize their response efforts.

3. **Configure channels**: Set up channels for alert notifications so your team is informed. You might simply use **SMS**, email, **Slack**, or **Discord**, or implement an incident management platform such as **PagerDuty**.

4. **Fine-tune thresholds**: When setting up alerts, you want to avoid alert fatigue. This will require you to fine-tune your alert thresholds. Overwhelming your team with unnecessary alerts will negate the efficacy of your alert system.

5. **Test**: Regularly test your alert setup to ensure it works. You can even conduct drills to simulate performance situations to verify your system's efficacy.

Once your alert system has been set up, you need to determine your response approach.

## Alert and incident response

It is important to establish a formal approach to incident response. When establishing your alert response schema, consider the following questions:

- Which alerts warrant a response?
- Who responds to which alerts?
- What internal communication is necessary?
- What external communication is required?
- What follow-up actions are necessary?

A structured approach to responding to alerts and incidents can help ensure that critical issues are appropriately addressed. This can lead to rapid resolution and minimum downtime.

Key components of an effective alert and incident response approach include the following:

- **Alert acknowledgment**: When an alert is received, acknowledge it promptly. You may establish maximum response times for your team to follow.
- **Assessment**: Investigate each alert to understand the root causes. The goal is not only to fix a current issue but also to prevent the issue from repeating in the future.
- **Execution of plans**: Ensure that your team follows predefined incident response plans through issue resolution. These plans should consist of documented step-by-step procedures for common issues and complex problems.
- **Communication**: Keep internal and external stakeholders informed about the incident status and resolution progress using predefined communication channels.
- **Documentation**: After resolving the incident, the root cause(s), resolution steps, and lessons learned should be documented. When appropriate, conduct a post-incident review with your team to identify improvement opportunities.

When we set up a robust and purposeful monitoring and alerting system, we significantly increase our ability to effectively maintain and improve the performance of our Java applications, even as they scale.

## Maintenance strategies

We need a strategy to maintain our application's code that goes beyond simply responding to system alerts. When we take a purposeful approach to code maintenance, we can ensure the sustained reliability, availability, and performance of our Java applications.

The main concept is to maintain a balance between scheduled maintenance and reactive maintenance. The table that follows provides insights into each approach and includes their advantages and best practices.

| | Scheduled maintenance | Reactive maintenance |
|---|---|---|
| Approach details | A planned approach to update code on scheduled intervals | Addressing issues as they arise |
| Advantages | • Predictable downtimes<br>• Reduced risks of failures<br>• Continual optimization | • Immediate issue resolution<br>• Requires fewer resources compared to the scheduled maintenance approach |
| Best practices | • Establish and follow the maintenance schedule<br>• Test changes during maintenance windows<br>• Communicate the plan to internal and external stakeholders | • Implement robust monitoring and alerting systems<br>• Create incident response plans<br>• Create documentation to support troubleshooting |

Table 16.3: Comparison of scheduled maintenance and reactive maintenance approaches

Once you have established your maintenance approach, you should consider documentation and knowledge management. Let's look at that in the next section.

## Documentation and knowledge management

Effective documentation and knowledge management is undeniably vital for maintaining a healthy code base. It can also help ensure smooth transitions during maintenance activities.

Documentation should be comprehensive, especially for large systems. Comprehensive documentation should consist of the following main components:

- API documentation
- Code documentation
- Configuration documentation

It is also important to have a knowledge-sharing component, which can include internal wikis and training for new team members. As systems are updated, the knowledge-sharing artifacts should be updated to ensure that they remain relevant. How knowledge is shared is almost as important as what is shared. Collaboration tools such as **Teams**, **SharePoint**, **Confluence**, and **GitHub wikis** can be used to facilitate collaborative document and knowledge sharing.

Maintenance strategies often include **refactoring**, which we will review in the next section.

## Refactoring strategies

Refactoring is the process of repurposing existing code without changing its external behavior. The purpose of refactoring is to improve the code's readability, maintainability, and performance.

We can identify the need for refactoring through **code smells**, which are signs of poorly designed code, duplicate code, and large or long classes and methods. Performance bottlenecks and complex logic are additional indicators that code should be refactored.

The following techniques can be used for refactoring your code:

- Break down large classes and methods into smaller, more manageable components.
- Rename classes, methods, and variables so they are meaningful and self-describing. This added code clarity improves overall readability and maintainability.
- Remove unused code.
- Simplify complex conditional logic into clearer, more readable constructs.

When refactoring our code, we should strive to refactor in small, incremental steps. This will minimize the risks of introducing new bugs. We should always write **unit tests** (see *Chapter 17, Unit and Performance Testing*) before and after refactoring. This helps us ensure that the functionality remains unchanged. Lastly, we should use a version control system to track our changes and support code **rollback** if necessary.

Sometimes we inherit systems and must maintain legacy code. That is covered in the next section.

## Legacy code

Legacy code refers to older code bases. They are typically difficult to maintain due to outdated practices, lack of documentation, or deprecated technologies (that is, systems coded with the **Common Business Oriented Language (COBOL)**).

When inheriting a legacy code base, we should first conduct a thorough code review to identify problematic areas, outdated dependencies, and security vulnerabilities. Armed with knowledge from the code review, we can prioritize critical sections of the code that require immediate attention or pose significant risks or security vulnerabilities.

When feasible, we should adopt a modernization strategy for the legacy code. This could include the following components:

- Implementing incremental updates
- Using code wrappers for new features

- Automating testing for changes
- Documenting legacy code and changes to it
- Implementing a backup schema
- Using a version control system
- Isolating legacy components from the rest of the system
- Training your developers on the legacy code and your maintenance plan

Implementing these maintenance strategies can help you ensure the ongoing stability and performance of your Java applications. The strategies are aimed at making the code easier to update, maintain, and scale over time.

## Summary

This chapter explored the essential practices and tools for effective code monitoring and maintenance, focusing on ensuring long-term performance, reliability, security, and scalability. We began with an overview of APM tools, detailing their key features such as real-time monitoring, transaction tracing, error tracking, and user experience monitoring. Popular APM tools for Java were reviewed, along with best practices for their implementation and use.

We emphasized the importance of code reviews for the goal of maintaining high-quality code. We covered best practices, detailed the peer review process, and shared common pitfalls, offering avoidance solutions.

The concept of logging was examined, starting with the fundamentals, including log levels, messages, and targets. We outlined best practices for effective logging, such as using appropriate levels, avoiding sensitive information, and centralizing logs. We also introduced popular logging frameworks for Java and discussed techniques for analyzing and managing log data.

Monitoring and alerting were highlighted. We covered how to set up a comprehensive monitoring system, as well as alert configuration and incident response strategies. We concluded the chapter by covering maintenance strategies. We compared scheduled and reactive maintenance approaches, stressing the importance of proactive planning. We emphasized the role of documentation and knowledge management in maintaining healthy code bases. Finally, we explored effective refactoring strategies and provided guidance on how to deal with legacy code.

In the next chapter, we will introduce strategies for creating and using unit and performance tests to help create and maintain high-performance Java applications. Specifically, we will look at unit testing, performance testing, and overarching strategies.

# 17

# Unit and Performance Testing

The importance of thoroughly testing our code cannot be overstated; moreover, this testing should be efficient. As our systems grow in complexity and scale, it becomes increasingly critical for us to ensure that every component of our software functions accurately and efficiently. This is where unit and performance testing comes into play. This is the focus of this chapter.

The chapter starts by covering **unit testing**, which we use to verify individual units of code. The goal is to ensure that the units perform efficiently and as expected. Through unit testing, we can catch anomalies (bugs) early, before the code is deployed into a production environment. Performance testing is introduced as a complementary process, whereby we test our software under various conditions to assess behaviors such as responsiveness, availability, scalability, reliability, and scalability. As the chapter demonstrates, performance testing can help us identify potential bottlenecks and ensure our systems can handle anticipated use cases and loads.

Both theoretical and hands-on approaches are taken in this chapter, giving you the opportunity to gain knowledge and experience with unit and performance testing. We will cover overarching strategies to include integrating both types of testing, automation, test environments, continuous testing, and feedback loops.

By the end of this chapter, you will have a thorough understanding of unit and performance testing and be able to leverage them to enhance the reliability and efficiency of your Java applications.

This chapter covers the following main topics:

- Unit testing
- Performance testing
- Overarching strategies

# Technical requirements

To follow the examples and instructions in this chapter, you will need the ability to load, edit, and run Java code. If you have not set up your development environment, refer to *Chapter 1*, *Peeking Inside the Java Virtual Machine (JVM)*.

The finished code for this chapter can be found here:

```
https://github.com/PacktPublishing/High-Performance-with-Java/tree/
main/Chapter17
```

# Unit testing

The software developer's adage of "test, test, and test again" still applies to modern systems but in a more refined way. Instead of testing our systems as a whole, we focus on small components of our code to ensure that they perform efficiently and as expected. These components are referred to as **units**. When we isolate units of code, we can more easily detect bugs and improve the overall quality of our code. This approach is referred to as unit testing and is the focus of this section.

> **Unit testing**
>
> Unit testing is an approach to software testing that involves testing the smallest sections of a system's code to ensure that it performs correctly and efficiently in isolation.

The primary benefits of unit testing include the following:

- **Bug detection**: Unit testing enables us to detect bugs early, before the code is published as part of the larger system.

- **Code quality**: This testing approach, with its finite focus, results in higher code quality.

- **Documentation**: The process of unit testing includes documentation of each unit's functionality, purpose, connectivity, and dependencies.

Now that we understand what unit testing is and why it is important, let's look at two popular unit testing frameworks.

## Frameworks

One of the most common unit testing frameworks is **JUnit**, perhaps because of its simplicity and ease of integration with **Integrated Development Environments (IDEs)**. Another popular framework is **TestNG**, which is comparatively more flexible and has functionality in addition to JUnit.

We will focus on JUnit and demonstrate how to write a unit test in the next section.

## Writing unit tests

There are several ways you can write and implement unit testing. Here is a straightforward approach:

1.  Ensure that you have a recent version of the **Java Development Kit (JDK)** installed.

2.  Download and install the **JUnit Jupiter API and Engine JARs**. The process for accomplishing this will depend on your IDE.

3.  Assuming that you are using Visual Studio Code, install the **Test Runner for Java** extension.

To demonstrate unit testing, we will write a simple calculator program, as shown here:

```java
public class CH17Calculator {
  public int add(int a, int b) {
    return a + b;
  }

  public int subtract(int a, int b) {
    return a - b;
  }

  public int multiply(int a, int b) {
    return a * b;
  }

  public int divide(int a, int b) {
    if (b == 0) {
      throw new IllegalArgumentException("Division by zero");
    }
    return a / b;
  }
}
```

As you can see, our application contains methods for adding, subtracting, multiplying, and dividing two numbers based on passed parameters. Next, let's create a `testing` class:

```java
import org.junit.jupiter.api.Assertions;
import org.junit.jupiter.api.BeforeEach;
import org.junit.jupiter.api.Test;

public class CH17CalculatorTest {

  private CH17Calculator calculator;
```

```
@BeforeEach
public void setUp() {
  calculator = new CH17Calculator();
}

@Test
public void testAdd() {
  int result = calculator.add(3, 4);
  Assertions.assertEquals(7, result);
}

@Test
public void testSubtract() {
  int result = calculator.subtract(10, 5);
  Assertions.assertEquals(5, result);
}

@Test
public void testMultiply() {
  int result = calculator.multiply(2, 3);
  Assertions.assertEquals(6, result);
}

@Test
public void testDivide() {
  int result = calculator.divide(8, 2);
  Assertions.assertEquals(4, result);
}

@Test
public void testDivideByZero() {
  Assertions.assertThrows(IllegalArgumentException.class, () -> {
    calculator.divide(1, 0);
  });
}
}
```

As you can see in the preceding code, we have created individual methods to test the methods in the primary class file. The next step is to run the test. Using Visual Studio Code, you can select the testing icon (the beaker in the leftmost panel). Now you can run individual tests, or all of the tests, by selecting the **Run Test** button to the right of each test name.

Figure 17.1 – The Visual Studio Code navigation panel

The test results will be available in the bottom section of your **Test Results** IDE interface. Here are the test results for the `testMultiple()` method:

```
%TESTC   1 v2
%TSTTREE2,CH17CalculatorTest,true,1,false,1,CH17CalculatorTest,,[engin
e:junit-jupiter]/[class:CH17CalculatorTest]
%TSTTREE3,testMultiply(CH17CalculatorTest),false,1,false,2,tes
tMultiply(),,[engine:junit-jupiter]/[class:CH17CalculatorTest]/
[method:testMultiply()]
%TESTS   3,testMultiply(CH17CalculatorTest)
%TESTE   3,testMultiply(CH17CalculatorTest)
%RUNTIME84
```

This demonstrates a simple unit test. Next, let's review some best practices to help us get the most out of unit testing.

## Best practices

A key best practice for unit testing is to keep the tests as small as possible. This will give us a tighter focus and allow us to more rapidly troubleshoot and resolve issues through code edits.

Another best practice is to ensure that our tests are **isolated tests**, which are tests that are independent of any external factors. This helps ensure that any errors or issues we detect are caused by our code and not an external environment. If we do not take this approach, we might struggle to efficiently determine the source of any errors.

A third practice is to ensure that our tests cover a myriad of scenarios, including error conditions and even edge cases. This is a thorough approach that strives to test for any possible situation. The extended time that this approach takes is worth it, as our approach can help ensure that our systems can perform under both routine and irregular conditions.

A fourth best practice, **assertions**, will be covered next.

### Assertions

Assertions are an important best practice. This best practice is simply to leverage assertions to validate expected outcomes.

> **Assertions**
>
> Assertions are code statements that are used to check whether a condition is true.

When an assertion fails, it indicates that the condition evaluated by the assertion is false, which usually results in the unit test failing. There are several assertion methods that we can use in JUnit. Let's look at four of the most common assertion methods.

**Method**: `assertEquals`

**Syntax**: `Assertions.assertEquals(expected, actual);`

**Example**:

```
int result = calculator.add(3, 6);
Assertions.assertEquals(9, result, "3 + 6 should equal 9");
```

**Method**: `assertNotEquals`

**Syntax**: `Assertions.assertNotEquals(unexpected, actual);`

**Example**:

```
int result = calculator.subtract(24, 8);
Assertions.assertNotEquals(10, result, "24 - 8 should not equal 10");
```

**Method**: `assertTrue`

**Syntax**: `Assertions.assertTrue(condition)`

**Example**:

```
boolean result = someCondition();
Assertions.assertTrue(result, "The condition should be true");
```

**Method**: `assertFalse`

**Syntax**: `Assertions.assertFalse(condition);`

**Example**:

```
boolean result = someCondition();
Assertions.assertFalse(result, "The condition should be false");
```

To use an assertion in our application, we simply add a line of code to each unit test. For example, as you review the `testAdd()` method in our `CH17CalculatorTest` application, you will see that it uses the `assertEquals()` assertion method:

```
public void testAdd() {
    int result = calculator.add(3, 4);
    Assertions.assertEquals(7, result);
}
```

Now that we have an understanding of some best practices in writing unit tests, let's review some common pitfalls.

## Pitfalls

As powerful as unit testing is, it also comes with several pitfalls. Here are three common pitfalls involved with unit testing and how to avoid them:

| Pitfall | Avoidance strategy |
|---------|--------------------|
| Ignoring edge cases | When we ignore edge cases in our unit testing, our systems can have undetected bugs. Ensure that you include a robust edge case strategy in your unit testing. |
| Over-testing | Over-testing in our context is creating tests that are too large, covering multiple units. To avoid this pitfall, create unit tests that are isolated from external dependencies and are focused on a single unit of code. |
| Under-testing | Under-testing refers to not running tests frequently enough to catch issues, especially when changing environments and scaling systems. To avoid this pitfall, perform tests frequently. |

Table 17.1 – Unit testing pitfalls and avoidance strategies

We will conclude our coverage of unit testing with a look at **Test-Driven Development (TTD)**.

## TDD

TTD is an interesting software development approach whereby the unit tests are written before the code is. The TDD cycle is often referred to as **Red-Green-Refactor** and is illustrated in *Figure 17.2*.

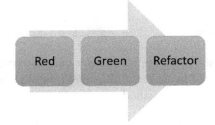

Figure 17.2 – The TTD cycle

TDD implementation starts with the Red step, wherein we write a test that fails because the related functionality has not been written. Next, in the Green step, we write the minimum amount of code required so the test will pass. Finally, in the Refactor step, we refactor our code to make it more efficient, improve its readability, and ensure that all related unit tests continue to pass.

The advantages of implementing TDD include the following:

- Promoting clean code
- Ensuring that the code is testable
- Helping developers thoroughly consider requirements before writing code

There are also a few challenges to the TDD approach, including the following:

- It is not beginner-friendly
- It requires a mental paradigm shift
- It can slow initial development

Now that we have a firm handle on unit testing, let's explore performance testing in the next section.

# Performance testing

The crux of this book has been to ensure our Java applications perform at the highest possible level. Several strategies, tools, and techniques have been presented to help us achieve our goal of high performance. In the previous section, we covered unit testing to help us ensure proper functionality. With performance testing, we will test our applications to see whether they can perform under various conditions and loads. This testing strategy involves evaluating the following characteristics of our applications: efficiency (speed), stability, responsiveness, reliability, and scalability.

There are several primary objectives of performance testing, including the determination of whether performance criteria have been met. We also want to identify performance bottlenecks so that we can refine our code. An additional objective is to ensure that our application can handle anticipated system and user loads.

## Types and tools

The five primary types of performance tests are detailed in the table that follows:

| Type | Focus |
| --- | --- |
| **Endurance testing** | Checks for memory leaks and resource depletion with a sustained load over extended time |
| **Load testing** | Tests performance with a specific number of concurrent users |
| **Scalability testing** | Checks scalability by adding transactions and users |
| **Spike testing** | Determines whether the application can handle a sudden increase in load |
| **Stress testing** | Pushes the load past capacity to determine the breaking point |

Table 17.2 – Performance test types

It is important to implement a performance testing plan that includes each type of performance test with an added focus on the types that are more critical to your specific application and goals.

There are several tools available to help us with performance testing. The three most common ones are featured in the table that follows. The table includes use cases for each:

| Tool | Description | Use case |
| --- | --- | --- |
| Apache Bench | Basic command-line tool for benchmarking HTTP servers | Simple load tests of HTTP services |
| Apache JMeter | Open source tool for load testing | Comprehensive testing with a variety of protocols (that is, HTTP, FTP, and so on) |
| Gatling | Advanced open source tool that simulates high user loads | Advanced load testing scenarios |

Table 17.3 – Performance testing tools

Let's wrap up this chapter with a look at big-picture approaches that are specific to unit testing and performance testing.

# Overarching strategies

This chapter's previous two sections were focused on unit testing and performance testing. This final section considers how we might combine the two types of testing for a cohesive strategy. This duality of testing is critical to our ability to develop and maintain robust and highly efficient Java applications.

We will start with a look at how to integrate the two types of testing.

## Integrating unit and performance testing

There are a few strategies we can adopt for incorporating both types of testing. The first approach is **parallel testing**, which involves running unit tests and performance tests in parallel. This approach can save us time. Another approach is **shared test cases**, which can make our testing more efficient. This approach allows us to leverage shared test data and potential configurations. A third, more advanced strategy is to use a **unified testing framework**. These frameworks support both types of testing and can ensure a seamless transition between them.

Regardless of our implementation approach, we want to ensure that we have comprehensive test coverage. To accomplish this, we should use tools to measure our code coverage for both testing types. This is referred to as **coverage analysis** and helps us ensure that all critical paths are tested. We should also use **incremental testing**, whereby we gradually increase our test coverage until all code has been covered by tests. Finally, we should conduct a **cross-validation** of our test results with performance outcomes. This validation is used to confirm that functionality is accurate and performance efficiency is acceptable.

# Summary

This chapter explored the critical role of testing in ensuring the reliability and efficiency of our Java applications. We started with an introduction to unit testing, highlighting its purpose, benefits, and best practices for writing effective tests. We then covered performance testing, explaining its objectives and the various types such as load and stress testing. The chapter concluded with a look at overarching strategies to integrate both testing types seamlessly into the development workflow, emphasizing the need for unified frameworks and comprehensive test coverage to enhance the overall quality and performance of our Java applications.

In the next chapter, we will take an extensive look at how to leverage **Artificial Intelligence** (**AI**) tools and technologies to help ensure that our applications are as efficient as possible and that they perform at the highest possible level. The chapter offers several opportunities for developers to harness the power of AI for the betterment of their Java applications' performance.

# 18

# Leveraging Artificial Intelligence (AI) for High-Performance Java Applications

The secret is out; **Artificial Intelligence (AI)** is here to stay and continues to revolutionize nearly every industry, including software development. AI is already a game changer and a success enabler. It is also already being leveraged to help developers significantly enhance the performance of their Java applications. This chapter aims to help Java developers understand and leverage AI to create and maintain high-performing Java applications.

The chapter starts with an introduction to AI in Java, covering its relevance to achieving high performance and discussing the current and future directions of AI. Next, the chapter provides a specific look at how AI can be used to optimize code, conduct performance tuning, and use **Machine Learning (ML)** models to predict performance bottlenecks.

Our coverage includes an exploration of integration with AI services and platforms. This includes insights into popular AI tools and best practices for achieving seamless integration. We would be remiss if we did not analyze the ethical and practical considerations regarding the use of AI for high-performance Java applications. Specifically, we will look at the ethical implications of using AI in software development and explore related practical challenges. Our coverage of ethical considerations includes ensuring fairness and transparency in AI-driven systems.

The main topics covered in this chapter are as follows:

- Introduction to AI in Java
- AI for performance optimization
- AI-powered monitoring and maintenance

- Integration with AI services and platforms
- Ethical and practical considerations

# Technical requirements

The finished code for this chapter can be found here:

```
https://github.com/PacktPublishing/High-Performance-with-Java/tree/
main/Chapter18
```

# Introduction to AI in Java

Software developers at all levels of experience have embraced AI as part of their workflows. AI-related tools and techniques can significantly enhance our ability to develop high-performing Java applications. There are two aspects for us to consider. First, we can use AI tools to help us develop, test, and enhance our code. Secondly, we can incorporate AI into our applications to introduce innovation and enhance functionality.

In this section, we will look at AI's relevance to high-performing Java applications, current trends in AI for Java, and future directions for AI.

## AI's relevance to high-performance Java applications

The power of AI technologies is undeniable. With ML, a subset of AI, we can train models to make advanced predictions, analyze data, and automate complex tasks. Java developers can leverage these capabilities to improve their development, testing, and maintenance of Java projects.

Here are four key methods by which we can leverage AI as part of our Java development and system support efforts:

- **Automated monitoring**: We can use AI to enhance legacy monitoring tools. This can lead to automated anomaly, bug, and bottleneck detection. The output is alerts that let us stay informed of our system's performance and provide targeted refactoring and optimization. The use of automated monitoring with AI can minimize downtime and safeguard the user experience.

- **Performance optimization**: AI can analyze large data sets quickly, which can be leveraged to help us identify bottlenecks. Moreover, AI can suggest optimizations based on its own analysis. Taking this a step further, we can use ML models to predict which parts of our code are most likely to cause issues. This lets us be proactive instead of reactive with our optimizations.

- **Predicative analytics**: AI and ML models are increasingly being used in various industries to make predictions based on dataset and trend analysis. For Java developers, this can take the form of forecasting future system loads and performance issues. This allows us to make informed decisions about infrastructure, scaling, and resource allocation.

- **Predictive maintenance**: Employing predictive maintenance models can help us accurately anticipate future failures, both in hardware and software. This allows us to plan, take preventative maintenance actions, prevent performance degradation, and improve the overall efficiency of our systems.

Now that we have a sense of how we can leverage AI for high-performance Java applications, let's look at some current trends related to AI in Java.

## Current trends

There are four AI-related trends that are shaping how Java developers integrate AI and harness its power:

- **Cloud services**: Each of the major cloud service providers (**Amazon Web Services (AWS)**, **Google Cloud**, and **Microsoft Azure**) has AI services that can be integrated with Java applications. The models vary among service providers and generally provide pre-built models and **Application Programming Interfaces (APIs)** for complex tasks such as **image recognition**, **predictive analytics**, and **Natural Language Processing (NLP)**.

- **Edge AI**: The concept of **edge computing** was realized with the wide adoption of cloud services. The concept is simply deploying systems and data close to the user to increase response times and reduce network latency. **Edge AI**, an extension of edge computing, involves deploying AI models on edge devices.

- **Libraries and frameworks**: There is an increasing number of AI-specific libraries and frameworks becoming available to Java developers. The goal of these libraries and frameworks is to simplify our implementation of AI models in Java applications. Notable libraries and frameworks worth researching include the following:

  - Apache Spark MLlib

  - **Deep Java Library (DJL)**

  - TensorFlow for Java

- **Transparency**: As our use of AI increases, the need to document how AI decisions are made are understandable. **Explainable AI (XAI)** calls for AI decisions and processes to be transparent.

The trends that we have described are currently seen in the industry. The next section reveals what we might see in the future.

## Future directions

The efficiencies and capabilities we can currently gain from the use of AI tools and technologies are impressive. It is exciting to consider which future directions these tools and technologies might take and how we might be able to leverage them to enhance our ability to create and maintain high-performing Java applications. Here are four future trends:

- **AI-driven development tools**: As AI tools and techniques become mature, they are likely to be built into our **integrated development environments (IDEs)**. AI-specific development tools, a new breed of IDEs, could emerge for our use in the next couple of years.

- **AI for cybersecurity**: Cybersecurity's importance increases as new AI tools and technologies are released. AI can be used to detect and even respond to cybersecurity threats to our systems. This capability is likely to increase in the coming years.

- **Hybrid models**: It is common for similar technologies to be introduced independently and then later combined to form a hybrid model. For example, **augmented reality** and **virtual reality** were introduced separately and later formed a hybrid model referred to as **mixed reality**. This is like with AI, ML, deep learning, and other related systems.

- **Quantum computing**: Quantum computing is a field that is ripe for the use of AI. The power of quantum computing's computing power married with the intelligence of AI stands to revolutionize AI and how we use it.

AI tools and technologies can enhance our high-performance Java toolkits. By understanding and leveraging AI technologies, we can create applications that are not only efficient and scalable but also intelligent and adaptive. In the next section, we will take a specific look at AI for performance optimization.

# AI for performance optimization

As we discussed in the previous section, AI represents an incredible opportunity for developers to enhance their performance optimization efforts and improve their results. By adopting a set of AI tools and models, we can efficiently improve our applications as part of a continuous improvement mindset. For example, the predictive abilities of AI can help us predict bottlenecks before they happen. Imagine updating your code to prevent a bottleneck or resource depletion without any system user being impacted. We no longer need to wait for complaints of low responsiveness or read logs to see errors and alerts.

Next, we will look at how we can use AI for code optimization and performance fine-tuning.

## Code optimization and performance tuning

AI is highly capable of analyzing existing code and providing optimization insights. This is possible by training ML models to identify code patterns that are inefficient, code that might lead to memory leaks, and additional performance issues.

Let's illustrate this with an example. The code that follows uses an inefficient way of data processing. This might not be noticed when monitoring your application but could become catastrophic if the data were to grow exponentially. Here is the inefficient code:

```java
import java.util.ArrayList;
import java.util.List;

public class CH18Example1 {
  public static void main(String[] args) {
    CH18Example1 example = new CH18Example1();
    example.processData();
  }
  public void processData() {
    List<Integer> data = new ArrayList<>();
    for (int i = 0; i < 1000000; i++) {
      data.add(i);
    }
    int sum = 0;
    for (Integer num : data) {
      sum += num;
    }
    System.out.println("Sum: " + sum);
  }
}
```

The AI-optimized code is provided next. A description of the optimizations follows the code:

```java
import java.util.stream.IntStream;

public class CH18Example2 {
  public static void main(String[] args) {
    CH18Example2 example = new CH18Example2();
    example.processData();
  }
  public void processData() {
    int sum = IntStream.range(0, 1000000).sum();
    System.out.println("Sum: " + sum);
  }
}
```

Using AI, the code has been optimized in three specific ways:

- The optimized code no longer uses an `ArrayList`. This is a smart change because that data structure can consume a lot of memory and increase the time it takes to process its contents.

- The code now uses an `IntStream.range` to generate the range of integers. It also computes the sum without the need to create an interim collection to store.

- Lastly, the optimized code uses Java's `IntStream` to efficiently handle the range of integers and perform the summation operation. Here, our code benefits from the inherent optimized nature of Streams' better performance.

AI was able to make three optimizations to our simple but inefficient code. Imagine what it could do for more robust applications.

Next, let's examine how ML models can predict issues such as bottlenecks.

## Predicting performance bottlenecks

ML models can be taught to predict performance bottlenecks. These models learn by ingesting historical performance data. From there, they can identify patterns that are likely to lead to a performance bottleneck. To demonstrate this, we will use the **Waikato Environment for Knowledge Analysis** (**WEKA**) platform.

> **WEKA**
> WEKA is an ML and data analysis platform. It is free software issued under the GNU General Public License.

Here is a simple ML example that we can use to predict performance bottlenecks:

```java
import weka.classifiers.Classifier;
import weka.core.Instances;
import weka.core.converters.ConverterUtils.DataSource;

public class CH18Example3 {
    public static void main(String[] args) throws Exception {
        DataSource source = new DataSource("path/to/performance_data.
        arff");
        Instances data = source.getDataSet();
        data.setClassIndex(data.numAttributes() - 1);

        Classifier classifier = (Classifier) weka.core.
        SerializationHelper.read("path/to/model.model");
        double prediction = classifier.classifyInstance(data.
        instance(0));

        System.out.println("Predicted Performance: " + prediction);
    }
}
```

We used the Weka library to load historical performance data and then, using a pre-trained `classifier` model, predict potential performance issues. Here is some simulated output we might expect from our code:

```
Predicted Performance: Bottleneck
```

There are a lot of assumptions that come with the preceding simulated output. For context, let's assume that we trained our Weka model to classify code as either `Normal` or `Bottleneck`.

Let's now move on to a review of a real-world case study.

## Abbreviated case study

This section provides an abbreviated case study to help illustrate the practical application of using AI for performance optimizations. The case study uses a Java-based web application scenario experiencing lag during peak usage times.

1. **Phase 1**: The first phase of this case study involved data collection. This included gathering specific performance metrics including response times, memory and CPU use, transaction rates, and more. This was collected by the development team and consisted of 12 months' worth of data.

2. **Phase 2**: Now that the development team had 12 months of performance data, they trained an ML model to predict potential performance bottlenecks. This training led to the model's ability to identify patterns in the data that led to performance degradation.

3. **Phase 3**: At this point, the development team had the data and the trained ML model. Next came the implementation phase, wherein they integrated the now-trained model into their application monitoring system. The system monitoring resulted in alerts and optimization scripts when a bottleneck was predicted. The optimized scripts were designed to adjust resource allocation and optimize database queries.

4. **Phase 4**: This final phase was the results phase. The development team had successfully collected data, then trained an ML model and implemented it in their monitoring system. The results were significant and included a 35% improvement in response times and a 42% decrease in slowdowns.

This case study demonstrates practical applications with tangible benefits of using AI to help us optimize performance.

Our next section will focus on leveraging the power of AI to help us monitor and maintain our code.

# AI-powered monitoring and maintenance

We covered the critical purpose of monitoring and maintenance extensively in *Chapter 16, Code Monitoring and Maintenance*. We can extend that topic beyond the manual interventions and predefined benchmarks and thresholds. AI-powered monitoring and maintenance presents a shift toward a proactive and efficient approach that uses ML and other AI techniques to detect anomalies, predict bottlenecks and failures, and automate responses.

This section looks at how we can leverage AI for anomaly detection, automated monitoring, logging, and alerting. We will also explore maintenance strategies that use predictive maintenance models. Let's get started with a look at anomaly detection using AI.

## Anomaly detection

One of the primary goals of monitoring is to detect anomalies that can lead to significant issues. AI's ability to ingest and analyze copious amounts of data empowers it to detect anomalies when the data is being reviewed by non-AI tools or humans.

The code that follows is an example Java application that uses an AI model for anomaly detection. In particular, this example looks at performance metrics:

```java
import org.deeplearning4j.nn.multilayer.MultiLayerNetwork;
import org.deeplearning4j.util.ModelSerializer;
import org.nd4j.linalg.api.ndarray.INDArray;
import org.nd4j.linalg.factory.Nd4j;

public class CH18Example4 {
    public static void main(String[] args) throws Exception {
        MultiLayerNetwork model = ModelSerializer.
        restoreMultiLayerNetwork("path/my_anomaly_model.zip");

        double[] performanceMetrics = {75.0, 85.7, 500, 150};
        INDArray input = Nd4j.create(performanceMetrics);

        INDArray output = model.output(input);
        double anomalyScore = output.getDouble(0);

        System.out.println("Anomaly Detection System Report (ADSR):
        Anomaly Score: " + anomalyScore);

        if (anomalyScore > 0.3) {
            System.out.println("Anomaly Detection System Report
            (ADSR): Anomaly detected!");
        } else {
            System.out.println("Anomaly Detection System Report
            (ADSR): System is operating normally.");
        }
    }
}
```

Our preceding example uses a neural network model to analyze performance metrics. The computation results in an anomaly score with output that varies based on the score. This can be used to help identify areas that we need to review or trigger an automated response.

Next, let's look at how AI can be used for logging and alerting.

## AI-based logging

As expected, we can use AI to enhance our logging and alerting systems. AI tools can provide us with more efficient and contextual alerts than otherwise possible. Let's look at a simple implementation of an AI logging system that includes alerting:

```
import org.apache.logging.log4j.LogManager;
import org.apache.logging.log4j.Logger;

public class CH18Example5 {
    private static final Logger logger = LogManager.
    getLogger(CH18Example5.class);

    public static void main(String[] args) {
        double cpuUsage = 80.0;
        double memoryUsage = 90.0;
        double responseTime = 600;

        if (isAnomalous(cpuUsage, memoryUsage, responseTime)) {
            logger.warn("Anomalous activity detected: CPU Usage:
            {}, Memory Usage: {}, Response Time: {}", cpuUsage,
            memoryUsage, responseTime);
        } else {
            logger.info("System operating normally.");
        }
    }

    private static boolean isAnomalous(double cpuUsage, double
    memoryUsage, double responseTime) {
        return cpuUsage > 75.0 && memoryUsage > 85.0 && responseTime >
        500;
    }
}
```

As you can see in the preceding example, the implementation is simplistic and results in notifications and warnings. It's noteworthy that this is simply provided for illustration purposes. In a full implementation, a pre-trained ML model would be incorporated to detect what is anomalous and what is not.

Let's end this section with an exploration of maintenance strategies with predictive maintenance models.

## Maintenance strategies

AI technology includes the ability for us to train models to predict maintenance. This model could be used to anticipate hardware and software failures long before they occur. When we are notified of the predictions, we can be proactive with our hardware and software maintenance.

Let's look at a simple program that uses AI to conduct predictive maintenance concerns:

```java
import weka.classifiers.Classifier;
import weka.core.Instances;
import weka.core.converters.ConverterUtils.DataSource;

public class CH18Example6 {
    public static void main(String[] args) throws Exception {
        DataSource source = new DataSource("path/my_maintenance_data.
        arff");
        Instances data = source.getDataSet();
        data.setClassIndex(data.numAttributes() - 1);

        Classifier classifier = (Classifier) weka.core.
        SerializationHelper.read("path/my_maintenance_model.model");
        double prediction = classifier.classifyInstance(data.
        instance(0));

        if (prediction == 1.0) {
            System.out.println("Predictive Maintenance System Report
            (PMSR): Maintenance required soon.");
        } else {
            System.out.println(" Predictive Maintenance System Report
            (PMSR): System is operating normally.");
        }
    }
}
```

Our example uses an ML model to predict whether maintenance is required. This is based on analysis of historical data.

As detailed in this section, AI can power our anomaly detection, automate our monitoring, enhance our logging and alerting systems, and predict when hardware and software maintenance is required. In the next section, we will briefly explore how we can integrate our Java applications with AI services and platforms.

# AI integration

AI requires significant computing power. Leveraging the elastic cloud computing of cloud service providers is commonplace. There are also open source platforms that offer pre-built models for use. Let's take a brief look at AI services from the three largest cloud service providers (AWS, Microsoft Azure, and Google Cloud) and two open source platforms (Apache Spark MLlib and TensorFlow for Java), starting with the three cloud service providers.

- **AWS AI Services**: Amazon's cloud platform offers an entire suite of AI services that include **SageMaker** for building training models, **Amazon Rekognition** for image and video analysis, and **Amazon Comprehend** for NLP.

- **Microsoft Azure AI**: Microsoft offers similar AI tools to AWS, which include **Azure ML** for ML model development, **Bot Service** for creating AI-powered chatbots, and **Cognitive Services**, containing pre-built AI functionality.

- **Google Cloud AI**: Google Cloud, like Amazon and Microsoft, offers a suite of AI tools. These include **AutoML**, which can be used for custom model training, **Vision AI** for image recognition, and **Natural Language API** for text analysis.

Now let's review two open source AI platforms:

- **Apache Spark MLlib**: This is a scalable ML library that is, as the name suggests, an add-on to Apache Spark. This library includes many algorithms that can be used for classification, clustering, collaborative filtering, and regression. These are all ripe for use with Java applications.

- **TensorFlow**: This is an open source library that focuses on numeric computation and ML. One of the great things about this library is that it provides **Java bindings** that enable us to use ML capabilities in our Java applications.

Let's look at best practices for integrating AI services into our Java applications.

## Best practices

Integrating AI services from one of the cloud service providers or an open source platform can seem daunting. The service providers have a plethora of documentation to aid in the implementation. Regardless of which platform or library you implement, the following best practices apply:

- Clearly define the problem you want to solve with AI. This clarity will help ensure that your implementation is efficient and purposeful.

- Ensure that the data you use to train your ML modes is clean and of high quality. The better your data is (that is, the more optimized its, quality, accuracy, and organization are), the more easily your ML models can learn from it.

- Optimize the performance of your AI operations. As previously stated, AI operations are computationally heavy, so optimization is critical.

- Just like other sophisticated software, AI services can fail (that is, produce unexpected results or crash), so be sure to incorporate robust error and exception handling.

- Be sure to continuously monitor the performance of your AI methods and modules. Maintain the code so that it remains optimized.

Next, we will review ethical and practical considerations regarding using AI in software development.

## Ethical and practical considerations

It is important to consider the ethical and practical implications of incorporating AI into our Java applications. It is easy to succumb to the power offered by AI in how it can help us significantly enhance the efficiency and performance of our applications. This should not overshadow the moral obligation to consider the challenges related to data privacy, fairness, and transparency.

Let's look at some of the key ethical implications of using AI in our applications. For each implication, a solution is suggested:

| Ethical implication | Challenge | Solution |
|---|---|---|
| **Data privacy and security** | AI models require large datasets and they can contain sensitive user information. | Implement data anonymization techniques. |
| **Fairness and bias** | AI models can unintentionally perpetuate biases. | Use a diverse dataset that is representative of a wide spectrum. |
| **Transparency** | Deep learning networks can thwart an underlying understanding. | Document and make AI model decisions transparent. |

Table 18.1: Ethical implications of using AI, with suggested solutions

Next, let's look at several practical challenges of using AI in our applications. For each challenge, a solution is suggested:

| Practical challenge | Challenge | Solution |
|---|---|---|
| **Continuous learning and maintenance** | AI models must be continuously updated. | Implement automated pipelines for retraining models. |
| **Model interpretability** | We need to understand how AI models make their decisions so that we can troubleshoot and debug. | Use interpretable models and document the model learning process. |

| Practical challenge | Challenge | Solution |
|---|---|---|
| **Performance overhead** | AI is computationally heavy. | Optimize AI models, breaking them down into smaller component modules. |
| **Scalability** | Scaling AI-powered applications can drastically increase overhead. | Design with scalability in mind. Use scalable frameworks. |

Table 18.2: Practical challenges of using AI, with suggested solutions

Lastly, let's review some strategic approaches to ensuring fairness and transparency in our AI-driven systems:

- **Clear documentation**: Documentation is a good practice for all software development and can be especially important when implementing AI models. Document the process, decisions, identified biases, strategies, limitations, and changes.

- **Regular audits**: Once you incorporate AI in your application, it is important to regularly conduct audits of your models. You should check for compliance with ethical standards and biases. These audits can be conducted manually or with the aid of automated tools.

- **Stakeholder involvement**: Internal and external stakeholders should be involved in the design, development, and deployment of AI models. Depending on the size of your organization, you might consider adding experts from the following areas:

  - Domain experts

  - Ethicists

  - User representatives

- **User education**: It should not have to be said that communicating how AI is used in applications is critical. This transparency builds trust and is simply the right thing to do.

Incorporating AI into our Java applications represents tremendous potential benefits. It also comes with ethical and practical challenges. By addressing these challenges proactively, we can create AI-driven systems that are not only high-performing but also fair, transparent, and trustworthy. This is a professional and unadulterated approach to AI use in our Java applications.

## Summary

This chapter explored how AI can be integrated into Java applications to enhance performance, efficiency, and reliability. The chapter covered an introduction to AI in Java, which included an overview of AI's relevance to high-performance Java applications, highlighting current trends and future directions in AI. We also examined how AI can be used for code optimization and predicting performance bottlenecks.

The chapter also examined how AI can help improve monitoring and maintenance processes through anomaly detection, AI-based logging, and alerting systems. We also looked at the concept of predictive maintenance models. A review of AI service platforms included a look at TensorFlow, Apache Spark MLlib, AWS AI Services, and Google Cloud AI.

We ended the chapter with a look at ethical and practical considerations, not as an afterthought to the book, but as a final important concept to leave you with. We addressed the ethical implications of using AI, including data privacy, fairness, transparency, and accountability. Solutions and best practices were discussed to ensure responsible and ethical AI integration.

By understanding and implementing the AI tools and techniques covered in this chapter, we are better able to create high-performance applications that are not only efficient and scalable but also ethical and trustworthy.

# Epilogue

You have reached the end of the book, and I hope that reading the chapters was a good use of your time. It is now important to consider high performance with Java from a holistic viewpoint and reflect on the core themes and insights explored throughout the book.

Our path to mastering Java high performance should be a continuous process of learning and trial and error, as well as a willingness to adapt to new challenges. The collection of concepts and techniques covered in this book was designed to help provide you with a solid foundation, but the real mastery will come when you start applying these concepts to your unique Java projects.

Java continues to evolve and the current tools and techniques for performance optimization are incredibly sophisticated. The insights that can be gained from understanding the Java Virtual Machine, optimizing data structures, fine-tuning memory management, and leveraging advanced concurrency strategies are invaluable for any serious Java developer. The book's exploration of frameworks, libraries, profiling tools, and emerging technologies such as AI highlights the fact that the dynamic nature of software development is core to our greater understanding. It also underscores the significance of the need to stay updated with the latest advancements.

The dedication to continuous improvement and performance excellence distinguishes a good developer from a great one. As you move forward, I encourage you to keep experimenting, stay curious, and never hesitate to dive deep into the inner workings of your code. Performance optimization is not just about writing faster code; it is about creating more efficient, scalable, and maintainable applications that can rise to meet both the demands of today and the challenges of tomorrow.

Thank you for joining me on this journey to achieve high performance with Java. I hope that the knowledge shared in these pages empowers you to build exceptional Java applications and inspires you to continue pushing the boundaries of what is possible.

# Index

packtpub.com

Subscribe to our online digital library for full access to over 7,000 books and videos, as well as industry leading tools to help you plan your personal development and advance your career. For more information, please visit our website.

## Why subscribe?

- Spend less time learning and more time coding with practical eBooks and Videos from over 4,000 industry professionals

- Improve your learning with Skill Plans built especially for you

- Get a free eBook or video every month

- Fully searchable for easy access to vital information

- Copy and paste, print, and bookmark content

Did you know that Packt offers eBook versions of every book published, with PDF and ePub files available? You can upgrade to the eBook version at packtpub.com and as a print book customer, you are entitled to a discount on the eBook copy. Get in touch with us at customercare@packtpub.com for more details.

At www.packtpub.com, you can also read a collection of free technical articles, sign up for a range of free newsletters, and receive exclusive discounts and offers on Packt books and eBooks.

# Other Books You May Enjoy

If you enjoyed this book, you may be interested in these other books by Packt:

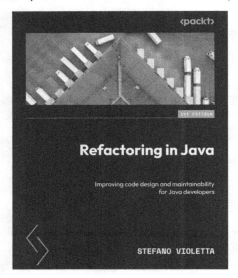

**Refactoring in Java**

Stefano Violetta

ISBN: 978-1-80512-663-8

- Recognize and address common issues in your code
- Find out how to determine which improvements are most important
- Implement techniques such as using polymorphism instead of conditions
- Efficiently leverage tools for streamlining refactoring processes
- Enhance code reliability through effective testing practices
- Develop the skills needed for clean and readable code presentation
- Get to grips with the tools you need for thorough code examination
- Apply best practices for a more efficient coding workflow

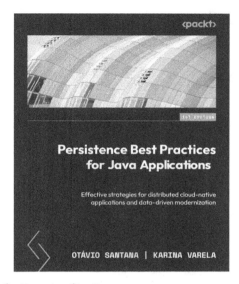

**Persistence Best Practices for Java Applications**

Otavio Santana, Karina Varela

ISBN: 978-1-83763-127-8

- Gain insights into data integration in Java services and the inner workings of frameworks
- Apply data design patterns to create a more readable and maintainable design system
- Understand the impact of design patterns on program performance
- Explore the role of cloud-native technologies in modern application persistence
- Optimize database schema designs and leverage indexing strategies for improved performance
- Implement proven strategies to handle data storage, retrieval, and management efficiently

## Packt is searching for authors like you

If you're interested in becoming an author for Packt, please visit `authors.packtpub.com` and apply today. We have worked with thousands of developers and tech professionals, just like you, to help them share their insight with the global tech community. You can make a general application, apply for a specific hot topic that we are recruiting an author for, or submit your own idea.

## Share Your Thoughts

Now you've finished *High Performance with Java*, we'd love to hear your thoughts! Scan the QR code below to go straight to the Amazon review page for this book and share your feedback or leave a review on the site that you purchased it from.

`https://packt.link/r/1835469736`

Your review is important to us and the tech community and will help us make sure we're delivering excellent quality content.

# Download a free PDF copy of this book

Thanks for purchasing this book!

Do you like to read on the go but are unable to carry your print books everywhere?

Is your eBook purchase not compatible with the device of your choice?

Don't worry, now with every Packt book you get a DRM-free PDF version of that book at no cost.

Read anywhere, any place, on any device. Search, copy, and paste code from your favorite technical books directly into your application.

The perks don't stop there, you can get exclusive access to discounts, newsletters, and great free content in your inbox daily

Follow these simple steps to get the benefits:

1.  Scan the QR code or visit the link below

https://packt.link/free-ebook/978-1-83546-973-6

2.  Submit your proof of purchase

3.  That's it! We'll send your free PDF and other benefits to your email directly

www.ingramcontent.com/pod-product-compliance
Lightning Source LLC
Chambersburg PA
CBHW080626060326
40690CB00021B/4835